DECONVOLUTION
OF ABSORPTION SPECTRA

DECONVOLUTION
OF ABSORPTION
SPECTRA

William E. Blass

George W. Halsey

Department of Physics and Astronomy
The University of Tennessee
Knoxville, Tennessee

1981

ACADEMIC PRESS

A Subsidiary of Harcourt Brace Jovanovich, Publishers

New York   London   Toronto   Sydney   San Francisco

CHEMISTRY

6659-1120

ACADEMIC PRESS, INC.
111 Fifth Avenue, New York, New York 10003

United Kingdom Edition published by
ACADEMIC PRESS, INC. (LONDON) LTD.
24/28 Oval Road, London NW1 7DX

Library of Congress Cataloging in Publication Data

Blass, W. E. (William E.)
   Deconvolution of absorption spectra.

   Bibliography: p.
   Includes index.
   1. Absorption spectra--Deconvolution. I. Halsey,
G. W. (George W.) II. Title.
QC454.A2B57   535.8'4   81-12667
ISBN 0-12-104650-8   AACR2

CONTENTS

PREFACE

This monograph deals with a rather universal concern—that of signal recovery. Signal recovery is a broad problem that spans numerous disciplines, and as a result there exists a rather rich general literature on the subject. A major problem results from the equally rich sets of technical terminology used in the literature.

Herein the reader will find a specialized aspect of signal recovery, namely deconvolution, discussed in the language of a physical scientist who, by accident or intent, has been crossbred with a sprinkling or more of the geophysicist, computer scientist, electrical engineer, and statistician.

Some readers will find the early chapters overly simplistic; we hope that they will find the later chapters on methods and results more stimulating. Other readers will find the early chapters quite demanding, and we hope that their background in the physical sciences will aid their entry into and advancement along the pathways of systems and signals.

To the extent that it is possible, we have attempted to produce an entry-level monograph for physical scientists which will enable them to begin tentative use of deconvolution methods in their work. Where elegance or a rigorous treatment has in our estimation been in conflict with effective communication, we have eschewed elegance or rigor or both.

ACKNOWLEDGMENTS

We should like to thank all of those who have shared with us their ideas and comments regarding deconvolution, especially Frank Grunthaner of the Jet Propulsion Laboratory and T. Harvey Edwards at Michigan State University. We also thank the past and present staff of the Molecular Spectroscopy Laboratory at the University of Tennessee, especially S. J. Daunt and D. E. Jennings. Without the broad-gauge support of the University of Tennessee Academic Affairs Office and the Vice-Chancellor for Academic Affairs, Walter Herndon, and without the policies regarding access to the extensive computer facilities of the university, the initial phases of our studies would have been impossible. Our special thanks go to the Head Professor of the Physics Department, W. M. Bugg, whose continuing tangible support and encouragement have made the later stages of this study possible. We further acknowledge and thank the secretarial staff of the Physics Department, who have patiently produced numerous revised versions of the typescript. We acknowledge support for portions of this effort provided by the Planetary Atmospheres program of NASA (grant NGL 43-001-006, principle investigator: N. M. Gailar). One of us (WEB) wishes to acknowledge the continuing support of Jacklynn Adkins Blass throughout the duration of the preparation of this monograph.

DECONVOLUTION
OF ABSORPTION SPECTRA

CHAPTER 1

INTRODUCTION

The world of experimental science is dependent on instruments that measure and record values of observable quantities. Many measurements are of a static variety, such as the length or mass of an object. The preponderance of experimental observations, however, involves the measurement of the magnitude of a quantity that varies as a function of a sequence-ordering variable that often quite naturally is taken to be time. Thus, the result of a particular experiment is not a single observation but a series of observations that vary with time.

It is worthwhile to digress from the mainstream of the discussion and consider the role of time as implied here. Many experimental observations actually do vary as a direct function of time viewed as a fundamental physical parameter. On the other hand, many observations, although not varying as a direct function of time as a driving independent variable, may quite naturally be thought of as a function of time. Time in these cases is viewed as a convenient sequencing or ordering variable. For example, as the diffraction grating of a dispersive spectrophotometer is rotated, the frequency of the radiation passed to the detector varies. The frequency of the detected radiation is physically representable as a function of the angle of incidence of the radiation on the diffraction grating relative to the normal to the grating.

Consider a single passed Littrow spectrometer [1–3]. The wave-number of the detected radiation is given by

$$\nu = nk \csc \theta \qquad (\text{cm}^{-1}), \tag{1}$$

where n is the order number, k is the reciprocal of twice the groove spacing, and θ is the angle of incidence of the radiation on the grating. The detected wavenumber is a function of θ. However, in the event that a scan is made from some starting angle θ_0 and the rate of change of θ is linear in time with a rate constant ω (rad/sec), the detected wavenumber may be characterized as

$$\nu = nk \csc(\theta_0 + \omega t), \tag{2}$$

and ν is seen to be a function of time. Such an extension is quite natural. If, however, the grating drive is powered by a stepping motor such that

$$\theta = \theta_0 + \Omega N_j, \tag{3}$$

where Ω is the change in θ per step and N_j is the step number of the jth observation, then Eq. (1) may be written as

$$\nu = nk \csc(\theta_0 + \Omega N_j). \tag{4}$$

Time does not appear in Eq. (4), nor does the stepping of the drive have to be linear in time. Nevertheless, we can consider N_j to be an ordering variable for the discrete observations of ν, and we can consider the sequence of ν values as a time series [4, 5] where time is viewed as a sequence-ordering variable whether or not it actually represents physical time.

Returning to the mainstream of the discussion, all experimental observations in which an instrument records a varying observable can be considered to yield a *time series* of values of the observable. Each measuring instrument, however, adds something of its own to the actual time-varying value of the sought-after observable. Eliminating, at least in part, that extra something that blurs our vision of the sought-after observable is the motivation for developing the *art* and *science* of deconvolution. For example, consider the vibration–rotation spectrum of a diatomic molecule. We generally think of such a spectrum as having a comblike appearance. The true spectrum at low pressures of the sample gas is composed of Gaussian-shaped spectral lines. Observing the spectrum with a spectrometer distorts (i.e., changes) the shape and width of the spectral lines, and if the spacing of the spectral lines is small enough compared to the resolution of the spectrometer, the comblike appearance will be distorted or even completely washed out. In this example, recovering the sought-after true spectrum—the comblike series of Gaussian lines—from the spectrometer output may be thought of as the goal of deconvolution.

In many experimental situations the desired data are representable as a function $L(\nu)$ and the output of the measurement system is representable as

$$S(t) = \int T(t - \nu)L(\nu)\,d\nu, \qquad (5)$$

where T is called the system transfer function. T also represents, in the general context here, the "something extra" that the measuring instrument (or system) adds to the time-varying value of the sought-after observable $L(\nu)$. We desire experimental data that characterizes $L(\nu)$, but we obtain $S(t)$ from our experimental measurement. Deconvolution is a method, when $T(t - \nu)$ can be measured or assumed, which allows us to obtain an estimator of $L(\nu)$ that, *if care is exercised*, can be quite certainly a better approximation to $L(\nu)$ than is $S(t)$ [6–12]. For example, in a spectroscopic experiment studying line shapes, the distortion of the exact transition profile by spectrometer and electronics is an undesirable complication. Deconvolution may permit one to obtain a better estimation of the true line profile.

In another instance, when a spectroscopic instrument is functioning near its theoretical resolution limit, and greater resolution is required, the only effective approach in the past has been to build a new, higher-resolution instrument (not always a feasible choice).

Now deconvolution techniques can be used to obtain a factor of 2–3 in resolution enhancement beyond the instrumental limit. That this is feasible shall be made reasonable as one proceeds through later chapters. Pliva, Pine, and Willson have recently demonstrated the achievement of a factor of 3–5 enhancement using data recorded with a signal-to-noise ratio of 1000:1 [13]. The future may see a greater utilization of deconvolution methods in order to recover previously unavailable desired data signals. In the field of image enhancement, such approaches are commonly used [12].

The highest-resolution spectroscopic measurements in the infrared are often limited in achievable resolution by detector signal-to-noise ratios. Opening the spectrometer slits to obtain a usable signal-to-noise ratio degrades resolution. A choice is then forced on the experimenter wherein he must make a trade-off between signal-to-noise ratio and resolution. Deconvolution should permit one to opt for a good signal-to-noise ratio and recover some of the "lost" resolution due to the wider entrance and exit slits. On the other hand, one might find that a factor of 2 resolution degradation allows expensive scientific systems to produce a significantly greater quantity of useful data. Deconvolution should permit lost resolution recovery with good confidence levels, thus resulting in more efficient use of state-of-the-art instrumentation.[†]

*Consideration of the ideas raised in the last paragraph can illustrate the more subtle aspects of deconvolution as an experimental tool. In order to elucidate this point, it is necessary to consider some relatively complex considerations regarding the acquisition of high-resolution infrared spectral data and then to apply these results to certain operational questions that arise in a high-resolution spectroscopy laboratory.

Let

S_0 = source intensity at some wavelength λ

α = the attenuation factor for slits of width W_0 [μm], α may be considered to be a function of grating order and angle

S_{W_0} = observed detector signal voltage at W_0 slit width.

Then

$$S_{W_0} = S_0 \alpha. \tag{6}$$

S_{W_0} is the equivalent signal voltage at the input to the amplifier for slits of W_0. Let β be the amplifier gain. Then for $\mathfrak{S}_S \equiv$ signal voltage at amplifier output, and $\mathfrak{S}_N \equiv$ noise voltage at the amplifier output,

$$\mathfrak{S}_S = \beta S_{W_0} = S_0 \alpha \beta \tag{7}$$

$$\mathfrak{S}_N = \beta N_{\tau_0}, \tag{8}$$

[†] The starred section may be skipped on a first reading (pp. 4–9).

where $N_{\tau_0} \equiv$ equivalent noise voltage at the amplifier input for a time constant equal to τ_0. (To be specific, we shall use an amplifier whose frequency passband $f_{pb} = 1/8\tau$.)

Defining the signal-to-noise ratio for slits of width W as Δ_W,

$$\Delta_{W_0} = \frac{\mathcal{G}_S}{\mathcal{G}_N} = \frac{S_0\alpha\beta}{\beta N_{\tau_0}} = \frac{S_0\alpha}{N_{\tau_0}} \tag{9}$$

at slits W_0 or

$$\Delta_{W_0} = \frac{S_{W_0}}{N_{\tau_0}}. \tag{10}$$

Let us choose a set of instrumental parameters, particularly slit width W_0 and time constant τ_0, that result in an acceptable signal-to-noise ratio, Δ_{W_0}, which we shall use as benchmark conditions in the following considerations.

Because the energy throughput of the spectrometer varies as the square of the slit width for equal width entrance and exit slits, we may write

$$S_W = S_{W_0}\left(\frac{W}{W_0}\right)^2. \tag{11}$$

In addition, since the signal-to-noise ratio varies as the square root of the amplifier time constant, we have

$$N_\tau = N_{\tau_0}\sqrt{\frac{\tau_0}{\tau}}. \tag{12}$$

The signal-to-noise ratio for slits W, Δ_W, is

$$\Delta_W = \frac{S_W}{N_\tau}. \tag{13}$$

Combining Eqs. (11) and (12) as specified by Eq. (13), it follows for a constant signal-to-noise ratio,

$$\frac{S_{W_0}}{N_{\tau_0}} = \frac{S_W}{N_\tau}, \tag{14}$$

that

$$\left(\frac{W}{W_0}\right)^2 = \sqrt{\frac{\tau_0}{\tau}} \tag{15}$$

or

$$\tau = \tau_0 \left(\frac{W_0}{W}\right)^4. \tag{16}$$

When changing the slit width from the benchmark value W_0 to a new value W, the time constant of the electronics system is governed by Eq. (16). That is, if we select a new slit width W, we must, in order to maintain a constant signal-to-noise ratio, select a time constant τ (and thereby a particular amplifier bandwidth) governed by Eq. (16).

The maximum scanning rate for a continuous scanning spectrometer has been discussed by Blass [14]. The following summarizes the treatment. Consider a spectral line to be a Gaussian of the form

$$\frac{1}{\sigma_\nu \sqrt{2\pi}} \exp\left[-\frac{(\nu - \nu_0)^2}{2\sigma_\nu^2} \right], \tag{17}$$

where ν_0 is the line center and σ_ν is related to the full width at half-maximum of the spectrum line (FWHM) $W_{1/2}$ as

$$W_{1/2} = 2.35\sigma_\nu. \tag{18}$$

Defining the scanning rate of a spectrometer as r_S (cm^{-1}/sec), we can specify the time to scan one FWHM (cm^{-1}) as

$$\tau_w = \frac{W_{1/2}}{r_S} = \frac{2.35\sigma}{r_S}. \tag{19}$$

It is a direct procedure to convert Eq. (17) to a true function of time:

$$\frac{2.35 r_S}{\sqrt{2\pi}\, W_{1/2}} \exp\left[-\frac{(t - t_0)^2}{2(W_{1/2}/2.35 r_S)^2} \right], \tag{20}$$

where t_0 in Eq. (20) corresponds to ν_0 in Eq. (17). Note that the scanning rate and FWHM of the spectral line appear in Eq. (20)

parametrically. Considering the power spectrum [15–17] of Eq. (20), Blass [14] was able to show that 99.5% of the spectral power in the spectrum line was contained below a frequency

$$f_M = \frac{2.35 r_S}{\pi W_{1/2}} = \frac{0.75 r_S}{W_{1/2}} \quad \text{(Hz)} \tag{21}$$

expressed as a function of scanning rate and FWHM of the spectral line. Thus, when scanning a spectrum at a rate r_S (cm^{-1}/sec) with an effective resolution of $W_{1/2}$ (cm^{-1}), the passband of the amplifying system f_{pb} must be at least

$$f_{pb} \geqslant f_M = \frac{0.75 r_S}{W_{1/2}}, \tag{22}$$

at which setting (assuming a single-stage RC filter with a 6 dB/octave roll-off) the voltage of 90% of the spectral components will be attenuated by less than 11% (1 dB or less).

Normally, one first chooses the passband (i.e., time constant) of the amplifier system that results in an acceptable signal-to-noise ratio at the slit width (resolution) desired. The scanning rate is then selected based on passband and resolution. From Eq. (22) we have that the scanning rate must satisfy

$$r_S \leqslant 1.33 f_{pb} W_{1/2} \quad (cm^{-1}/sec), \tag{23}$$

where

$f_{pb} = 1/8\tau$ (assumed herein, characteristic of several popular lock-in amplifiers)

$W_{1/2} =$ full width at half maximum of the instrumental resolution function.

so that

$$r_S \leqslant \frac{1.33}{8\tau} W_{1/2}. \tag{24}$$

The spectral slit width can be represented as a function of wavenumber, grating angle, monochromator focal length, and grating order for any particular monochromator design. We shall use the representation specific to a double-passed Littrow monochromator (27):

$$W_{1/2} \approx \Delta\nu = \frac{W\nu}{2Fm} \cot\theta, \qquad (25)$$

where

ν = monochromator center frequency (cm^{-1})
θ = angle of incidence
W = slit width
F = focal length of main collimator
m = number of passes of grating.

Defining

$$\psi \equiv \frac{\nu \cot\theta}{2Fm}, \qquad (26)$$

we can write

$$W_{1/2} = \psi W, \qquad (27)$$

and the scanning rate can be expressed as

$$r_S = \frac{1.33\psi W}{8\tau}. \qquad (28)$$

Let r_{S_0} be the maximum scanning rate at the so-called benchmark condition and r_S, the scanning rate at a different slit width (thus at a different resolution). It follows that

$$r_{S_1} \leqslant \frac{1.33}{8\tau_1} \psi W_1, \qquad (29)$$

$$r_{S_0} \leqslant \frac{1.33}{8\tau_0} \psi W_0. \qquad (30)$$

So that

$$\frac{r_{S_1}}{r_{S_0}} \leqslant \left(\frac{W_1}{W_0}\right)\left(\frac{\tau_0}{\tau_1}\right). \qquad (31)$$

Using the results obtained in Eq. (16) to eliminate τ_0 and τ from Eq. (31), we find

$$\frac{r_{S_1}}{r_{S_0}} \leqslant \left(\frac{W_1}{W_0} \right)^5. \tag{32}$$

That is, the maximum permissible scanning rates vary as the fifth power of the ratio of the slit widths and thus as the fifth power of the ratio of the resolution.*

Consider the case where Δ, the signal-to-noise ratio, is determined from the best possible instrumental resolution case, i.e., the situation where one is pushing instrumental capabilities and just achieving a manageable signal-to-noise ratio for the narrowest possible slits W_0 and largest reasonable time constant τ_0. The maximum scanning rate under these conditions is labeled r_{S_0}. Further, consider the impact of instrumental resolution degraduation on scanning rate (and thereby on total instrument time required to perform a given experiment). The results are presented in Table I.

If one can use deconvolution techniques reliably to recover a factor of 2.8 in resolution, it then follows that the data may be acquired 172.1 times faster than without using deconvolution tech-

TABLE I

Effect on Scanning Rate of Resolution Degradation for Constant Signal-to-Noise Ratio[a]

Resolution degradation factor	Maximum scan rate (units of r_{S_0})
1.0	1.0
1.2	2.5
1.5	7.6
1.7	14.2
2.0	32.0
2.5	97.7
2.8	172.1
3.0	243.0

[a]Benchmark conditions are specified by r_{S_0} at slit width W_0 and time constant τ_0.

niques. There are, of course, ridiculous cases where a factor of 172 means little at all. However, consider the case where best information infers that a single data run would take one year without deconvolution. Degrading the resolution by a factor of 2.8 (opening the slits to $2.8W_0$) and deconvolving the experimental data to recover the lost resolution implies an acquisition time (spectrometer utilization time) of slightly over two days plus computer deconvolution time (which will be a small quantity, though computer dependent). It is apparent that even though the initial conclusion usually drawn relative to the value of deconvolution revolves around resolution enhancement beyond practical instrumental capabilities, it is also true that the technique allows for the more efficient utilization of instrumental capability while achieving the same resolution as the practical limit of the instrument.

In some cases, deconvolution makes it possible to enhance instrumental throughput by large factors without any practical loss of information in the data. Consider the process of obtaining a spectrum of ν_3 of CH_4 at 3000 cm^{-1}. The Doppler width of the transitions is given by

$$W_D = 7.16 \times 10^{-7} \nu \left(\frac{T}{M} \right)^{1/2} (cm^{-1}), \qquad (33)$$

where ν is the frequency in cm^{-1}, T the absolute temperature, and M the molecular weight in AMU. In the preceding case, the FWHM (full width at half maximum) of an isolated spectrum line due to the Doppler effect at room temperature is 0.0093 cm^{-1}. Assume that a 100 cm^{-1} wide scan at an instrumental resolution of 0.008 cm^{-1} can be made in 120 hr; the scan will produce lines with an apparent resolution of about 0.012 cm^{-1}. On the other hand, the slits could be doubled, degrading resolution to 0.016 cm^{-1}, increasing the acquisition rate by a factor of 32 [cf. Eq. (32)]. The apparent resolution (roughly the square root of the sum of the squares of the Doppler width and the resolution width) would be 0.0185 cm^{-1}. Deconvolution by a factor of 2 would recover the 0.0093 cm^{-1} Doppler limited spectrum. The total instrumental time would be 120 hr/32, or 3.75 hr, as opposed to 120 hr. Additionally, the actual resolution would be better in the deconvolved data than in the instrumentally limited but nondeconvolved data.

Deconvolution is a technique that gives the experimental scientist additional capability and freedom in carrying out any given experimental program. In some cases it may allow highly specialized, one-of-a-kind research instruments to be more widely and fully utilized by passing some of the data acquisition burden to a computer system where development costs have been spread over many identical computer installations (cf. this type of discussion applied to computer control systems in Ref. [1]).

Noise, of course, complicates matters, as we shall see in the latter sections of this work. At this point we make a single assertion that should be kept in mind. Noise filtering can be viewed as the selective attenuation of high spectral frequencies (or in some cases low or specific frequencies), whereas deconvolution involves *restoration* of high spectral frequencies attenuated in the measurement process. The two processes are diametrically opposed, and thus a primal requirement for deconvolution is high signal-to-noise ratios in the data to be deconvolved.

The "solution" of the integral equation

$$S(t) = \int T(t - \nu)L(\nu)\,d\nu \tag{34}$$

for $L(\nu)$ is, essentially, the goal of deconvolution; there have been many investigations in several fields that are related to this general problem. More correctly, however, deconvolution is the process of recovering $L(\nu)$ from $S(t)$ knowing $T(t - \nu)$. Since Eq. (34) is the convolution of L and T as

$$S = L \oplus T = T \oplus L, \tag{35}$$

where $\oplus$ represents the convolution operation, one may take the Fourier transform of Eq. (35) and obtain

$$s = lt \tag{36}$$

and thus

$$l = \frac{s}{t}. \tag{37}$$

On retransforming l, one obtains L. It seems the problem is solved.

Unfortunately, except in certain cases where extremely favorable signal-to-noise ratios are available and the system transfer function $T(t - v)$ has certain desirable but generally unattainable characteristics, the Fourier transform solution to the problem, though analytically correct, is often plagued by application problems. It turns out that recovery of $L(v)$ by operations in signal space (as opposed to transform space) has often been found to be more reliable (cf. discussions in Ref. [12]).

This work is based on the Jansson *et al.* [8, 9, 11] modification of Van Cittert's [18] deconvolution method. The Jansson effort was directed toward resolution enhancement of high-resolution infrared spectra. A development of Van Cittert's techniques in the ESCA field is due to Wertheim [6, 19, 20]. Willson and Edwards have pursued tests and modifications of the Jansson method with good results [10].

The present work grew out of a desire to test the Jansson modification of the Van Cittert deconvolution method in order to ascertain something about the reliability of deconvoluted high-resolution infrared spectral data. A state-of-the-art resolution infrared spectrometer such as the one at the University of Tennessee is ideal for tests of deconvolution methods. One makes a data run at highest instrumental resolution, repeats it at half the resolution, repeats it at one quarter the resolution, etc. Then with measured instrument functions corresponding to each data set, one deconvolves the degraded resolution data sets and compares them to high-resolution, actual data records. Examples of such trials will be found in later chapters. The results are encouraging and demonstrate that deconvolution can be a reliable tool. We further set out to minimize required computation time and have succeeded in that regard to a large extent (cf. also Ref. [10] and [13]).

This work is not the final word in deconvolution—far from it. It is hoped that it will serve as a jumping-off point for scientists who wish to experiment with and use deconvolution methods in their own work.

As usual, nothing is "cheap" and so the deconvolution users can expect to invest a number of man-hours and a number of computer hours on deconvolution in their own laboratories.

CHAPTER 2

PHYSICAL
MEASUREMENT
SYSTEMS: OVERVIEW

The quantitative study of the behavior of submicroscopic physical systems invariably depends on probing the physical system with photons or electrons or other "beams" or "fields" and sensing, with instrumentation of varying complexity, the effects of the interaction of the physical system being studied and the usually well-defined "probe." Physical theory will often characterize the "signal" that is expected because of the interaction of the probe and the physical system under investigation. We shall call this signal (or its *idealized* detection and measurement) the *true* signal.

For example, consider a hypothetical two-state system shown in Fig. 1. We shall "probe" an ensemble of N of these two-state systems, prepared so as to maximize the probability that any member of the undisturbed ensemble is in the lower energy state. Because of the finite lifetime of a member of the ensemble in the excited state as well as because of the interaction among members of the ensemble, the energy levels are actually of finite width, as illustrated in Fig. 1. We shall probe the system with a beam of particles that interact with

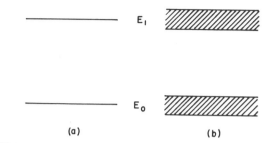

FIG. 1 Hypothetical two-state system: (a) idealized and (b) actual.

members of the ensemble by inelastic collisions resulting in popula-
tion of the excited state. The excited members of the ensemble decay
to the lower state, emitting a photon. The true signal produced by the
probe–physical system interaction *might* be characterized by physical
theory as $L(\nu)$ shown in Fig. 2, where A is a normalization factor.

The physical scientist is interested in the determination of $L(\nu)$.
Although this statement seems at first to be intuitively factual, on
reflection one might decide that a particular experiment demands
only the determination of, say, ν_0 and σ (relative to the hypothetical
example and Fig. 2). Considering the state-of-the-art recovery methods for $L(\nu)$, it is fortunate that often the determination of ν_0 and σ
(or like parameters) has been sufficient. However, more and more
often there are pressing reasons why $L(\nu)$ must be recovered from an
experiment, and we currently beg the question of necessity and simply
assume that it is important to recover the precise form of $L(\nu)$, the
desired output of a particular experiment.

Lest the reader be led astray, let us emphasize the fact that
although in this example $L(\nu)$ is analytically characterized (herein

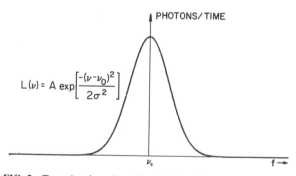

$$L(\nu) = A \exp\left[\frac{-(\nu - \nu_0)^2}{2\sigma^2}\right]$$

FIG. 2 True signal produced by the "probe"–system interaction.

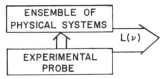

FIG. 3 Generalized representation of a physical system–"probe" interaction and the desired true output signal $L(\nu)$.

represented as a Gaussian profile), such is not generally the case. That is, the true signal is $L(\nu)$, but one generally cannot or does not wish to assume the actual analytical form of the profile. To do so at the signal recovery stage often builds a real bias into the results of the most fundamental experiments.

Proceeding in a more general fashion, let us represent all physical systems under investigation as in Fig. 3, which intentionally leaves the origin of the desired true signal $L(\nu)$ undefined. It may be the emission from the ensemble or the residual probe beam.

Again let us emphasize that $L(\nu)$ is the desired result of the physical experiment. $L(\nu)$ is seldom, if ever, the output signal of any realizable experimental measurement system. In general, the actual situation is represented in Fig. 4. $M(\nu')$ is the actual output of the physical measurement system. [For example, $M(\nu')$ might be the number of photons of frequency ν' per unit area per unit time at the detector of an optical spectrometer.] In the event that the experimental measurement system is characterizable as a linear system, the output $M(\nu')$ is related to the input $L(\nu)$ as

$$M(\nu') = \int H(\nu' - \nu)L(\nu)\,d\nu \qquad (38)$$

or

$$M = H \oplus L, \qquad (39)$$

where $\oplus$ represents convolution. $H(\nu' - \nu)$ completely represents the effect of the experimental measurement system on the true, physically

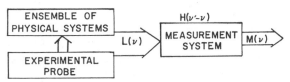

FIG. 4 Representation of an actual experimental system.

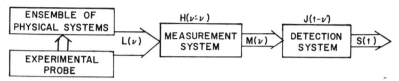

FIG. 5 Representation of an actual experimental measurement system including signal-processing subsystem.

meaningful signal $L(\nu)$. $H(\nu' - \nu)$ is called, among other names, the experimental measurement system transfer function.

Often $M(\nu')$ is not directly output by the experimental measurement system but must be detected by a sensor of some sort, which is then amplified, filtered, sometimes transformed, and recorded as represented in Fig. 5. $S(t)$ is the recorded system output signal. The effect of the detection, amplification, and recording system on $M(\nu')$ is representable as

$$S(t) = \int J(t - \nu')M(\nu')\,d\nu' \qquad (40)$$

or

$$S = J \oplus M. \qquad (41)$$

Notice that we have converted ν' to t to indicate that the output is often converted to a time series $S(t)$ with ν' and t usually related by some straightforward linear transformation.

Thus, the total effect of the experimental system on $L(\nu)$ is representable as

$$S(t) = J(t - \nu') \oplus H(\nu' - \nu) \oplus L(\nu), \qquad (42)$$

or replacing $J(t - \nu') \oplus H(\nu' - \nu)$ by $T(t - \nu)$, which we shall call the system transfer function or system instrument function,

$$S(t) = T(t - \nu) \oplus L(\nu). \qquad (43)$$

The effect of an instrumental measurement process is the convolution of the system transfer function with the true, and often implicitly desired, signal. Deconvolution is the process of recovering an estimator of the true signal from the instrumental output signal—the "undoing" of the convolution to the extent that this is possible. In the parlance of the signal theorist, deconvolution can be viewed as inverse

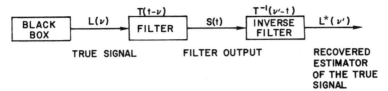

FIG. 6   Deconvolution from the signal theorist's viewpoint.

filtering [21]. Figure 6 represents deconvolution from the signal theorists's viewpoint. The recovered estimate $L^*(v'')$ of the true signal $L(v)$ is given by

$$L^*(v'') = L(v) \oplus T(t - v) \oplus T^{-1}(v'' - t). \tag{44}$$

Since complete recovery is not always possible, even in theory, the recovered signal is represented as $L^*(v'')$ rather than $L(v)$. The assumption of total recovery would require that

$$\delta(v - v'') = T(t - v) \oplus T^{-1}(v'' - t), \tag{45}$$

where $\delta(v - v'')$ is the Kronecker delta function.

Let us assume that a mechanism exists for determining an actual inverse transfer function, i.e., an estimator of T^{-1}. Call the inverse transfer function $F(v'' - t)$. Then

$$D(v'' - v) = F(v'' - t) \oplus T(t - v) \tag{46}$$

and the recovered signal $L^* (v'')$ is given by

$$\begin{aligned} L^*(v'') &= D(v'' - v) \oplus L(v), \\ L^*(v'') &= F(v'' - t) \oplus J(t - v') \oplus H(v' - v) \oplus L(v), \end{aligned} \tag{47}$$

where the total distortion of $L(v)$ is specified by $D(v'' - v)$. In fact, if the entire signal-processing chain is considered to be included in the experimental measurement system, $D(v'' - v)$ becomes the effective instrumental transfer function as represented in Fig. 7.

Since $L(v)$ is the actual desired information-carrying signal, when one speaks of a transfer function it would be most appropriate to speak of $D(v'' - v)$. However, it is not always possible to determine

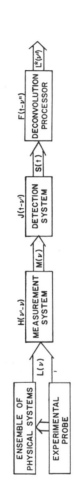

FIG. 7 Representation of the effective instrumental transfer function $D(\nu' - \nu)$, the convolution of $H(\nu' - \nu)$, $J(t - \nu')$, and $F(t - \nu'')$.

$F(\nu'' - t)$, especially when the best deconvolution process is an iterative, data-dependent process. In fact, the successful use of deconvolution techniques to achieve resolution enhancements of a factor of 2 or 3 precludes the use of inverse filtering discussed previously. Signal space deconvolution *as discussed in the following chapters* is an intrinsically nonlinear process, whereas inverse filtering as described earlier is an intrinsically linear process. *It is the nonlinearity of the signal space deconvolution process that allows one to achieve significant resolution enhancements* [12].

The following deals with one aspect of the problem of deconvolution. We treat the recovery of $L^*(\nu)$ from $S(t)$ by an iterative process in signal space. We do not treat the actual determination of $F(\nu'' - t)$ and thereby $D(\nu - \nu'')$, the effective instrumental transfer function depicted in Fig. 7. The utility of strict inverse filtering is limited [12] and the method has been discussed by Halsey and Blass [1, 21].

CHAPTER 3

PHYSICAL MEASUREMENT SYSTEMS: AN APPLICATION

In order to make the rather abstract systems concepts of the previous chapter real and concrete for the reader, we herein apply the concepts to optical spectroscopy, and specifically to high-resolution, infrared absorption dispersive spectrophotometry [2, 3].

The physical system is an ensemble of gas phase molecules, the probe, a "white light" beam of infrared photons. The signal $L(\nu)$ is the depletion of the probe beam photons resulting from molecular absorption as a function of wavelength λ or wave number $(1/\lambda)$. (Since wavenumber times the speed of light is frequency, we choose for convenience to speak of wavenumber or frequency as frequency. Where a distinction is important, it is noted.) The probe beam is not generally of constant intensity as a function of frequency. However, assuming a constant intensity probe beam over some region of interest simplifies this example and can actually be a reasonable assumption in practice. Figure 8 represents the situation pictorially,

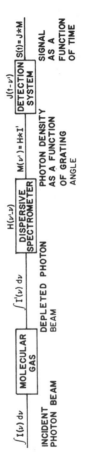

FIG. 8 Dispersive spectrometer signal evolution.

where the integral over $I(\nu)$ represents the total input probe beam and over $I'(\nu)$, the depleted beam. The function $I'(\nu)$ is identifiable as $L(\nu)$ in Chapter 2. Prior to the dispersive spectrometer in Fig. 8, however, $L(\nu)$ is not available, only its integral over frequency. Thus, the dispersive spectrometer, and specifically the monochromator, separates out each frequency of radiation as a function of some experimentally controlled variable such as angle of incidence of the photon beam on the diffraction grating, as indicated in Fig. 8. Following the angular dispersion of the photon beam as a function of photon frequency, the signal $M(\nu')$, the photon density as a function of frequency is available at the detector position. We have chosen to define $M(\nu')$ prior to detection, since for the optical spectrometer, as well as other experimental measurement systems, the choice of detection methods is quite distinct from the actual experimental measurement system. Choosing $M(\nu')$ as the detector output voltage (or current) would not change anything in a significant fashion. $M(\nu')$ is then detected by a photoconductive or photovoltaic detector which with associated bias circuitry produces a voltage signal referred to grating incidence angle and thereby as a function of photon frequency. Typically, this signal is amplified and synchronously detected by a lock-in amplifier. To a large extent most dispersive spectrometric systems have common features to this point in the signal-processing chain. Two choices are possible at this point: analog recording of the signal $S(t)$ on a chart recorder or digital sampling of $S(t)$ by a computer control system.

Since analog recording of the signal $S(t)$ by and large precludes further signal processing unless the record is digitally encoded at a later time, we shall only treat the digital recording option. Typically, the signal $S(t)$ is sampled by a sample-and-hold amplifier and the sampled signal passed to an analog-to-digital (A/D) converter, which then provides the digital equivalent of the analog voltage signal to the computer for storage on a bulk storage medium such as magnetic tape or disk for further processing [2, 22].

There are some often neglected but unusual things that "happen" to the desired signals as they pass through the system depicted in Fig. 8. Figure 9 represents a view of the signal production and storage chain. Remember, $I(\nu)$ and $I'(\nu) = L(\nu)$ are actually not accessible as a function of frequency before processing by the dispersive spectrometer, even though we have chosen in Fig. 9 to display these signals as a function of frequency. The most important aspect presented in

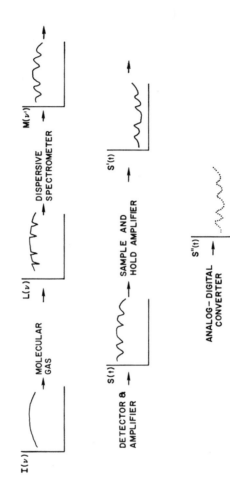

FIG. 9 Dispersive spectrometer signal production and storage sequence.

Fig. 9 is the inversion and offsetting of the signal produced by the sample-and-hold amplifier. Careful analysis shows that no significant difference results if $I'(v)$ is assumed to be inverted and offset as long as $I(v)$ does not vary significantly over the frequency range of a particular spectral scan. One must, however, note the fact that the sample-and-hold amplifier (in this example) generates a constant offset equal to full-scale output of the lock-in amplifier. If $I(v)$ ($\approx$ constant) produces, say, a 6-V output from the lock-in amplifier while full-scale output is 10 V, the signal $S''(t)$ will contain a 4-V bias that must be removed before processing the digitally recorded record.

If one were to implement a real time digital filtering and deconvolution program as a part of the digital sampling and recording system, then the total system depicted in Fig. 7 would have been established. As noted earlier, such a system does not often work well, and since it is a linear system does not enable optimum recovery of information.

CHAPTER 4

SIGNAL SPACE AND
FREQUENCY SPACE

From physical measurement systems similar to those described in the earlier chapters come data. Quite often the physical scientist thinks of his or her data as an amplitude record that is a function of frequency. It is, however, necessary to reorient one's thinking so that the data record [and the desired data signal $L(\nu)$] is seen as a series of data values as a function of time—a so-called time series [4, 5, 15–17]. Time here need not be time, but only a sequence-generating set of numbers, although a great deal of data are recorded linearly in time and so the required change in orientation should be straightforward.

Using the specific example of a Gaussian-shaped absorption line normalized to unit area, i.e.,

$$L(\nu) = \frac{1}{\sigma_\nu\sqrt{2\pi}} \exp\left[-\frac{(\nu - \nu_0)^2}{2\sigma_\nu^2} \right] \tag{48}$$

and the parameters $W_{1/2}$, the full width at half maximum (in cm^{-1}), and the spectrometer scanning rate r_s (in cm^{-1}/sec), we can convert

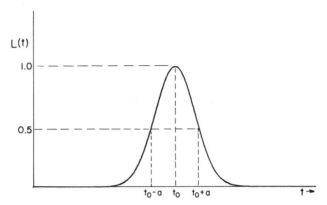

FIG. 10   Sample data signal as a function of time, where $a = W_{1/2}/2r_s$.

$L(\nu)$ to a true function of time [14]:

$$\hat{L}(t) = \frac{2.35 r_s}{\sqrt{2\pi}\, W_{1/2}} \exp\left[-\frac{(t-t_0)^2}{2(W_{1/2}/2.35 r_s)^2}\right]. \qquad (49)$$

$\hat{L}(t)$ is displayed in Fig. 10 and is the time series equivalent of a Gaussian spectral line with the FWHM (full width at half maximum) equal to $W_{1/2}$ cm^{-1}. The Fourier transform of $\hat{L}(t)$ is

$$L(\omega) = \exp\left[-\frac{(W_{1/2}\omega)^2}{2(2.35 r_s)^2}\right]\exp(-jwt_0), \qquad (50)$$

and the power spectrum of the spectral line $\hat{L}(t)$ is given by

$$|L(\omega)|^2 = \exp\left[-\frac{(W_{1/2}\omega)^2}{(2.35 r_s)^2}\right] \qquad (51)$$

and is displayed in Fig. 11. It can be shown [14] that 99.5% of the spectral power is contained between $\omega = \pm(4.70 r_s / W_{1/2})$ (radian/sec) or $(0.75 r_s / W_{1/2})$(Hz). Notice that, as expected, the narrower in cm^{-1} the spectral feature, the greater the maximum frequency content of the power spectrum of the time series $\hat{L}(t)$.

Let us assume that $\hat{L}(t)$ is the desired data to be measured and processed by the generalized system described in Chapter 2 and specifically by the spectrometric system described in Chapter 3.

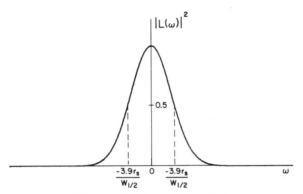

FIG. 11   Power spectrum of $L(t)$.

The spectral line $L(\nu)$ has a FWHM of $W_{1/2}$ (cm^{-1}), and this is equivalent to a FWHM in seconds for $\hat{L}(t)$ of $W_{1/2}/r_s$, where r_s is the scanning rate of the spectrometer in cm^{-1}/sec. Then 99.5% of the power spectral density of the time series (the data viewed in time) lies in the range $-0.75 r_s/W_{1/2} \leqslant f \leqslant 0.75 r_s/W_{1/2}$ [Hz].

If we assume that the experimental measurement system (the spectrometer) has a transfer function $H(\nu' - \nu)$, or equivalently, $\hat{H}(t' - t)$, then the output is $M(\nu')$ or $\hat{M}(t')$, respectively. For simplicity assume that $H(\nu' - \nu)$ is of a Gaussian form and has a FWHM of $R_{1/2}$, a measure of the monochromator resolution. The time equivalent is $R_{1/2}/r_s$. The output of the spectrometer at the detector position is

$$M(\nu') = H(\nu' - \nu) \oplus L(\nu), \tag{52}$$

or

$$\hat{M}(t') = \hat{H}(t' - t) \oplus \hat{L}(t). \tag{53}$$

Taking the Fourier transform of Eq. (53) we find

$$M(\omega') = H(\omega' - \omega)L(\omega) \tag{54}$$

and the power spectrum of $M(\omega')$

$$|M(\omega')|^2 = |H(\omega' - \omega)|^2 |L(\omega)|^2. \tag{55}$$

It is obvious from Eq. (55) that if the instrumental resolution (in time) $R_{1/2}/r_s$ is wider than the FWHM of the spectral data $W_{1/2}/r_s$, that

$|M(\omega')|^2$ will require a lesser frequency range to include 99.5% of its spectral power than will $|L(\omega)|^2$. These results are presented pictorially in Fig. 12. Also included is a representation of $\hat{M}(t')$, now noticeably broadened by the spectrometer [or equivalently, in frequency space, by the attenuation of the high-frequency components of $L(\omega)$ and thus of $\hat{L}(t)$]. It is worth noting that although the spectrometer (physical measurement system) can reduce the range of the spectral frequency content of the desired data $\hat{L}(t)$, the desired data are not accessible without processing of $\int L(\nu)\,d\nu$ by the spectrometer. Generally, the experimentalist thinks of the finite resolving power of the measurement system as limiting access to the desired data, and so it does. Herein we are suggesting a slightly different viewpoint, that the measurement system has a limited frequency passband for the desired data signal viewed as a time series.

The detection, amplification, and recording system also has a frequency passband determined by Fourier transforming the subsystem transfer function. Often the frequency passband of this portion of a system is under control of the experimenter, and so it needs only to be made broad enough to limit further distortion of the desired data $L(\nu)$ viewed as the time series $\hat{L}(t)$.

Consider the following example. Assume that you have a *perfect* high-fidelity recording of a symphony orchestra. (This is actually impossible, of course, but for the sake of the example assume that you possess such a masterpiece.) Further assume that you possess a perfect turntable and tone arm equipped with a perfect stylus and cartridge. Playing the recording, you have available a signal -10 mV $\leqslant \hat{L}(t) \leqslant +10$ mV at the cartridge output, and by our assumption $\hat{L}(t)$ is perfection itself and that the power spectral density of $\hat{L}(t)$ is such that $0 \leqslant |L(\omega/2\pi)|^2 \leqslant 100{,}000$ Hz. Yet your amplifier has a flat passband from 15 Hz to 20,000 Hz and rolls off on both ends to zero gain at 6 Hz on the low side and 28,000 Hz on the high side. Or if you will, the square modulus of the Fourier transform of the transfer function of the amplifier might be as shown in Fig. 13a. In addition, your speaker has a power spectrum response as shown in Fig. 13b. The power spectrum of the system transfer function is the product of Fig. 13a and 13b and is shown in Fig. 13c.

The points to notice are these: The amplifier permanently eliminates all frequencies above 28 kHz. There is no way that any of these frequencies can ever be recovered even if the amplifier output were recorded on the same magic machine that produced the perfect symphony recording.

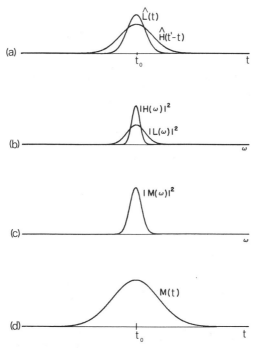

FIG. 12   Reduction in spectral frequency content as the data signal passes through the measurement system.

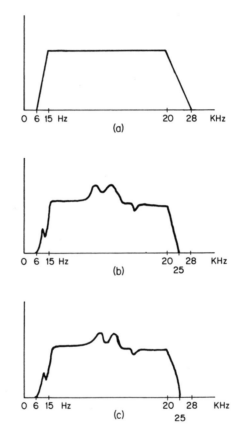

FIG. 13 Hypothetical power spectrum of (a) a high-fidelity amplifier, (b) a speaker, and (c) the complete system.

Consider the speaker with its typically uneven responses and resonances. With the exception of the complete attenuation (i.e., elimination, of frequencies above 25 kHz), all the "irregularities" could, in principle, be compensated for if a perfect recording were made of the system output.

To grasp deconvolution—a process that will be accomplished in signal space—one must see the desired data [i.e., $L(\nu)$] all the way back in the production chain as a time series (cf. Chapter 1) and begin to think of the effects of instrumentation on the power spectrum of the data time series.

Imagine a total system transfer function, $T(\nu, t)$ in Fig. 7, of the form of a sinc function $[(\sin x)/x]$ that has a Fourier transform as shown in Fig. 14a. If the power spectrum of the data time series $\hat{L}(t)$ [corresponding to $L(\nu)$] extends beyond $\pm f_m$, processing by the measurement system loses, for all time, data features generated by spectral frequencies above $|f_m|$.

If the power spectrum of the transfer function is as shown in Fig. 14b (Fourier transform of sinc^2), then again data features dependent on frequencies $0 \leqslant f \leqslant |\pm f_m|$ are attenuated and, in principle, may be recovered by the deconvolution process.

If the power spectrum is a Gaussian as shown in Fig. 14c, no finite frequencies are *in principle* ever completely lost and so should be able

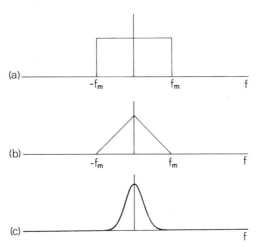

FIG. 14 Power spectra of three idealized system transfer functions of (a) a sinc function, (b) a sinc^2 function, (c) a gaussian function.

to be recovered by a deconvolution process. The presence of noise limits the generality of the last statement.

Thus deconvolution may be viewed as a power spectrum restoration process and the availability of inexpensive computational facilities has made development of such restoration methods both feasible and desirable. However, before succumbing to euphoria, it is well to note that the presence of noise in the data and/or measurement and processing system places severe constraints on the total restoration of the data power spectrum.

Up to this point we have been concerned with frequencies in the power spectrum. However, in considering the effects of noise on the deconvolution process, we must also consider signal and noise amplitudes as a function of spectral frequency. Figure 15 represents the power spectrum of a Gaussian data signal, the attenuation of high-frequency components by a narrower passband instrument with a Gaussian transfer function, and the relative amplitudes of signal and noise present at the output. The restoration of signal spectral frequencies significantly greater than f_n is not possible because of the presence of the noise.

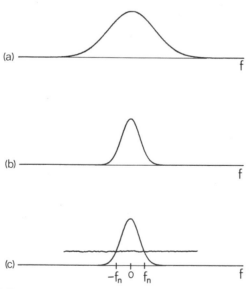

FIG. 15 (a) Power spectrum of a Gaussian data signal, (b) power spectrum of the output signal limited by the instrumental frequency passband, and (c) power spectrum of the output signal and the noise present in the instrumental output signal.

CHAPTER 5

DECONVOLUTION IN
SIGNAL SPACE

The object of deconvolution, in the abstract, is the recovery of the best estimation of $L(v)$ given $S(t)$ and knowing $T(t - v)$ in the convolution equation

$$S(t) = L(v) \oplus T(t - v), \qquad (56)$$

that is,

$$S(t) = \int_{-\infty}^{\infty} L(v) T(t - v) \, dv. \qquad (57)$$

There are, in fact, many different deconvolution techniques. Frieden [12] discusses each of many techniques in an extensive review of deconvolution from the point of view of image enhancement. Of special interest are those papers by Frieden and Biraud listed in the Bibliography. The work of both Frieden and Biraud is directly comparable to that discussed in this monograph. Their approaches are considerably different yet the results are of comparable quality, especially in regard to resolution enhancement.

Frieden's review is directly useful in spectroscopic studies, since the entire review is cast in one dimension instead of the usual two of picture processing. The following chapters specialize to the application of one particular signal space deconvolution technique that has been found to be very useful in resolution enhancement in the field of linear absorption spectroscopy.

Van Cittert [18] devised an iterative scheme for recovery of $L(\nu)$ using operations in signal space (as opposed to transform or frequency space). Briefly, the Van Cittert method used the relation

$$L_n = L_{n-1} + (S - L_{n-1} \oplus T) \tag{58}$$

and the starting assumption that

$$L_1 = S. \tag{59}$$

Historically, attempts to use the Van Cittert method met with varying degrees of success, for reasons discussed at the end of this chapter.

Jansson et al. [8, 9] developed a modification of the Van Cittert method. The major modification consisted of multiplying the correction to the L_{n-1} iterative estimator of the true spectrum by a relaxation parameter $\alpha_n(y)$ described below. Willson [10] later improved on the Jansson et al. techniques while he was working in the laboratory of T. H. Edwards at Michigan State University. In the following when we speak of the Jansson algorithm we are in fact speaking of the Jansson algorithm [8, 9] as modified by Willson [10]. This algorithm (called the Jansson algorithm in the following chapters) results in Eq. (58) becoming

$$L_n = L_{n-1} + \alpha_n(y)(S - L_{n-1} \oplus T), \tag{60}$$

where $y(t)$ is the amplitude of $L_{n-1}(t)$. Jansson's $\alpha_n(y)$ is shown in Fig. 16. Selection of $\alpha_{n_{max}}$ less than one, for example, uses only a partial correction in Eq. (60). The fact that $\alpha_n(y)$ falls off to zero at $y(t) = 0$ and $y(t) = y_{max}$ constrains the resulting nth approximation to L, i.e., L_n, such that L_n is usually positive or zero in all cases (assuming S was originally positive for all t) and limits the amplitude of $L_n(t)$ to a value y_{max}. Since Jansson was dealing with infrared absorption spectra that have no negative amplitudes and are limited to 100% absorption on the other end of the absorption scale, this function $\alpha_n(y)$ builds in desired constraints. In practice, because of the possibility that negative amplitudes can be generated within the Jansson algorithm with Jansson's (as well as other) relaxation parame-

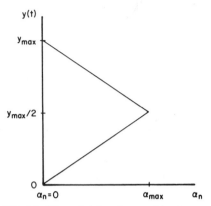

FIG. 16 The $\alpha_n(y)$ function used by Jansson.

ter, it is generally fitting to set all negative amplitudes to zero at the end of each iteration cycle.

The function $\alpha_n(y)$ is the key to the success of the Jansson algorithm. All the parameters in Eq. (60) represent various sequences of values and, were it not for the algebraic complexity of the resulting equation, would be written as $L_n(t)$, $L_{n-1}(t)$, $\alpha_n[y(t)]$, $S(t)$, and $T(t,t')$. Equation (60) represents the rule for constructing, say $L_n(5)$ from the $n-1$ iteration value for $L_{n-1}(5)$ corrected by the quantity to the right of the plus sign in Eq. (60). The quantity in parenthesis, which forms the complete correction in the Van Cittert scheme [Eq. (58)] is multiplied by the relaxation parameter $\alpha_n[y(5)]$. The quantity $y(5)$ is the amplitude of the $n-1$ iterative estimate of L at $t=5$, i.e., $L_{n-1}(5)$. Thus, the correction that is made is data dependent with the minimum (or zero) correction being made for very large and very small data amplitudes. It is the continually changing data dependence of the algorithm that results in its successful application. The data dependence also makes the Jansson algorithm a nonlinear process and as a result complicates the process of comprehending precisely how the algorithm works.

Before proceeding with more practical matters, we shall look at the Jansson and Van Cittert algorithms as well as a further modification of these algorithms in order to develop a picture of what is happening as we deconvolve. Recall the Van Cittert algorithm

$$L_n = L_{n-1} + (S - L_{n-1} \oplus T) \tag{61}$$

and the Jansson algorithm

$$L_n = L_{n-1} + \alpha_n(y)(S - L_{n-1} \oplus T). \tag{62}$$

We shall consider a modified algorithm of the form

$$L_n = L_{n-1} + \alpha_n(S - L_{n-1} \oplus T), \tag{63}$$

where α_n does not depend on $y(t)$. It can be shown rather directly that Eq. (63) may be written as

$$L_n = S \oplus \sum_{i=0}^{n-1} (-1)^i \left[\binom{n-1}{i} \beta + \binom{n-1}{i+1} \right] \left(\prod_{p=n-i}^{n} \alpha_p \right) T^i \tag{64}$$

or if all α_j are identical

$$L_n = S \oplus \sum_{i=0}^{n=1} (-1)^i \left[\binom{n-1}{i} \beta + \binom{n-1}{i+1} \right] \alpha^{i+1} T^i, \tag{65}$$

where for Eqs. (64) and (65)

$$T^i \equiv T \oplus T \oplus \cdots \oplus T \tag{66}$$

such that

$$T^3 \equiv T \oplus T \oplus T \tag{67}$$

and $\binom{n}{m}$ are binomial expansion coefficients where

$$\binom{n}{m} = \frac{n!}{(n-m)! \, m!} \tag{68}$$

and

$$L_1 = \alpha_1 \beta S \tag{69}$$

is assumed. Most investigators choose $\beta = 1/\alpha_1$, but for the purposes of symmetry especially, we retain the generalized starting assumption given in Eq. (69).

Thus, the Van Cittert algorithm and the modified algorithm of Eq. (63) are seen to be representable as the convolution of the observed spectrum with a "deconvolution filter" F_n given by everything to the right of the convolution operation in Eqs. (64) and (65). Jansson's algorithm is not directly representable as the convolution of the observed spectrum with a deconvolution filter due to the data dependence of $\alpha_n(y)$.

While valuable for calculations, Eqs. (64) and (65) are not especially helpful in seeing what happens in the Van Cittert or modified deconvolution. If, however, we begin with Eq. (63) [the Van Cittert algorithm is contained in Eq. (63) if $\alpha_i = 1$] and define the function I such that

$$I \oplus A = A \tag{70}$$

for any A, we can rewrite Eq. (63) as

$$L_n = \alpha_n S + L_{n-1} \oplus (I - \alpha_n T) \tag{71}$$

or

$$L_n = \alpha_n S + \alpha_n L_{n-1} \oplus \left(\frac{I}{\alpha_n} - T \right). \tag{72}$$

Let

$$G_n \equiv \left(\frac{I}{\alpha_1} - T \right) \oplus \left(\frac{I}{\alpha_2} - T \right) \oplus \cdots \oplus \left(\frac{I}{\alpha_n} - T \right) \tag{73}$$

and

$$G_0 \equiv I. \tag{74}$$

It follows directly that

$$L_n = S \oplus \left[\sum_{i=0}^{n-2} \left\{ \left(\prod_{p=n-i}^{n} \alpha_p \right) G_i \right\} + \left(\prod_{j=1}^{n} \alpha_j \right) \beta G_{n-1} \right] \tag{75}$$

and if $\alpha_i = \alpha$

$$L_n = S \oplus \left[\left\{ \sum_{i=0}^{n-2} \alpha^{i+1} G_i \right\} + \alpha^n \beta G_{n-1} \right]. \tag{76}$$

If $\alpha_1 = 1$ and $\beta = 1$, the standard starting assumptions, Eq. (75) becomes

$$L_n = S \oplus \left[\sum_{i=0}^{n-1} \left(\prod_{p=n-i}^{n} \alpha_p \right) G_i \right] \tag{77}$$

and if $\alpha_i = \alpha$, Eq. (77) becomes

$$L_n = S \oplus \sum_{i=0}^{n-1} \alpha^{i+1} G_i. \tag{78}$$

For $\alpha_i = 1$ we obtain the equivalent statement of the Van Cittert algorithm from Eq. (78),

$$L_n = S \oplus \sum_{i=0}^{n-1} G_i. \tag{79}$$

In Eqs. (77–79) we have a statement of the modified algorithm and the Van Cittert algorithm under the normal initial assumptions, i.e., $\alpha_1 = 1$, $\beta = 1$. If we define everything to the right of the convolution operator in Eqs. (77–79) as F_n (under the various assumptions specified), we can write

$$L_n = S \oplus F_n, \tag{80}$$

where F_n can be viewed as a deconvolution filter, with

$$F_n = \sum_{i=0}^{n-1} \left(\prod_{p=n-1}^{n} \alpha_p \right) G_i \qquad \text{modified} \tag{81}$$

$$F_n = \sum_{i=0}^{n-1} \alpha^{i+1} G_i \qquad \text{modified } \alpha_j = \alpha \tag{82}$$

$$F_n = \sum_{i=0}^{n-1} G_i \qquad \text{Van Cittert.} \tag{83}$$

Thus, the fifth approximation to L is given by

$$L_5 = S \oplus F_5. \tag{84}$$

The Jansson algorithm, as mentioned earlier, is not representable as the convolution of a filter with the observed spectrum, since the Jansson algorithm is essentially a nonlinear process due to the dependence of $\alpha_n(y)$ on the amplitude of the signal in the immediately preceding approximation to the desired signal. This must not be viewed as a disadvantage. It is, in fact, the nonlinear nature of the Jansson algorithm that allows one to successfully deconvolve and obtain significant resolution enhancement in a predictable fashion.

Before proceeding to the application of the Jansson algorithm, it is instructive to consider the limitations of the Van Cittert and modified algorithms following an analysis due to Frieden [12].

Recalling Eq. (78) for the modified algorithm, we can write

$$L_n = S \oplus \sum_{i=0}^{n} \alpha^{i+1} G_i, \tag{85}$$

where G_i is given by Eq. (73). Fourier transforming, one obtains

$$\hat{l}_n = \hat{S} \sum_{i=0}^{n} \alpha (1 - \alpha \hat{t})^i, \tag{86}$$

which may be analytically summed and yields

$$\hat{l}_n = \left(\frac{\hat{S}}{\hat{t}} \right) \left[1 - (1 - \alpha \hat{t})^{n+1} \right]. \tag{87}$$

This result, which is true at any spectral frequency in transform space, shows (after Frieden [12]) that

1. In the limit as $n \to \infty$, L_n approaches the inverse filtering estimate $(\hat{S}/\hat{t})$ if $|(1 - \alpha \hat{t})| < 1$. Thus, if $\alpha \hat{t}$ is negative due to noise, then $\hat{l}_n$ will not converge for that frequency and, more disastrous, neither will its transform $L_n(t)$.

2. For finite n, $[1 - (1 - \alpha \hat{t})^{n+1}]$ acts as a smoothing window in transform space and thus the modified algorithm provides a smoothed inverse filtering estimate. As n increases, the bracketed function approaches a Rect function.

3. As $\hat{t} \to 0$, Frieden shows that at frequencies for which $\hat{t} = 0$, the estimate in transform space becomes $\hat{l}_n = (n + 1)\hat{N}(\omega)$, where $\hat{N}(\omega)$ is the transform of the noise in the data. That is, $\hat{l}_n$ becomes simply a linearly enhanced version of the noise.

4. For small $\hat{t}$ and moderate n, the estimate given by Eq. (87) is just a linear version, $(n + 1)\hat{S}$, of the experimental data.

To apply the preceding to the Van Cittert algorithm, set $\alpha = 1$. It is easy to see that the modified algorithm makes it possible to avoid some difficulties with the Van Cittert algorithm by judicious selection of α_n. Mostly, however, we have found that this is only postponing the inevitable move to the Jansson algorithm. We feel that fully grasping the structure and function of the linear methods will assist researchers in the application of the much more useful Jansson algorithm.

CHAPTER 6

APPLYING THE JANSSON ALGORITHM

We have examined deconvolution in signal space in Chapter 5 and have seen that the fundamentally linear methods of solving Eq. (57) for $L(\nu)$ are simply signal space "equivalents" of the inverse filtering estimates obtained by Eq. (37). None of the preceding proves that the Jansson algorithm in Eq. (60) is superior, only that it is a nonlinear method. As pointed out by Frieden [12] in discussing image enhancement and restoration, only inherently nonlinear methods seem to yield continually fruitful returns.

This has been our experience with the Jansson algorithm—that it is capable of producing consistently good results when properly applied.

The Jansson algorithm is specified by Eq. (60), i.e.,

$$L_n = L_{n-1} + \alpha_n(y)(S - L_{n-1} \oplus T), \qquad (60)$$

where S is the original data record viewed as a time series, T is the system transfer function, and $L_n(\nu)$, the nth estimate of $L(\nu)$. Recall that $S(t)$ is the result of a measurement process representable as a convolution (cf. Chapter 1)

$$S(t) = \int_{-\infty}^{\infty} L(v) T(t - v) \, dv. \tag{88}$$

For the sake of simplicity, we shall assume that the numerical relationship between v and t is known and available and henceforth work with time series exclusively so that Eq. (88) formally becomes

$$S(t) = \int_{-\infty}^{\infty} L(t') T(t - t') \, dt'. \tag{89}$$

For example, let $L(t')$ be represented as

$$L(t') = \frac{1}{\sqrt{2\pi}\, a} \exp\left[-\frac{(t' - t_0)^2}{2a^2} \right] \tag{90}$$

and $T(t' - t)$ be represented as

$$T(t', t) = \frac{1}{\sqrt{2\pi}\, b} \exp\left[-\frac{(t' - t)^2}{2b^2} \right]. \tag{91}$$

Then $S(t)$ may be found from Eq. (89) to be

$$S(t) = \frac{1}{\sqrt{2\pi}\, (a^2 + b^2)^{1/2}} \exp\left[-\frac{(t - t_0)^2}{2(a^2 + b^2)} \right], \tag{92}$$

which provides an example of the application of Eq. (89). Equation (92) reflects the well-known result that the convolution of two Gaussian functions is also a Gaussian function.

Step by step the Jansson algorithm may be expressed as follows:

A. Set $j = 1$, $L_0(t) = S(t)$.
B. Form $L_j(t) = L_{j-1}(t) + \alpha_j(y)(S(t) - \sum_{t'} L_{j-1}(t') T(t - t'))$ where
    (i) y represents the value of $L_{j-1}(t)$ for the particular t at which $L_j(t)$ is being calculated;
    (ii) $S(t)$ is the observed time series;
    (iii) $L_{j-1}(t)$ is the previous estimate of $L(t)$;
    (iv) The sum over t' is over the range of nonzero values of $T(t - t')$.

    C. Apply convergence/termination tests. If satisfactory, the current L_j is the estimator of $L(t)$ and the procedure is completed. If tests are not satisfactory, set $j = j + 1$ and return to step B.

On exiting from the preceding iterative procedure an estimator of the desired true data set $L(t)$ has been found. Notice that at each step, $\alpha_j(y)(S - L_{j-1} \oplus T)$ is the current correction to the previous iteration. Forming

$$E_j = \left[\frac{1}{N} \sum_{\substack{\text{all points}}}^{N} \left[\alpha_j(y)(S - L_{j-1} \oplus T) \right]^2 \right]^{1/2}, \qquad (93)$$

essentially the rms correction to the previous estimator of $L(t)$ yields some idea of convergence of the deconvolution process. Figure 17 shows the typical behavior of E_j using the Jansson algorithm. We should emphasize that E_j is a measure of the *average* point-by-point correction. It is not an infallible measure of convergence, nor does a small E_j imply that the deconvolution process has achieved the determination of "truth." It is a useful measure of the progress of convergence; it does not determine whether, only how rapidly. The function $\alpha_n(y)$ in Eq. (60) is the factor that results in the Jansson algorithm being an inherently nonlinear process. There is a measure of freedom in the specification of this function. In general, the function used forces the known or desired data constraints into the process. For absorption spectra, which has been scaled to be in the range from 0 to 1, for example, Jansson [9] used

$$\alpha_n(y) = \alpha_n(L_{n-1}(t)) = \alpha_{n\,\max}\left[1 - 2|L_{n-1}(t) - \tfrac{1}{2}| \right],$$
$$\alpha_{n\,\max} = \text{const} \qquad (94)$$

which is shown in Fig. 16.

    Willson [10] used, instead of Jansson's triangular function, a quadratic function

$$\alpha_n(L_{n-1}(t)) = 4\alpha_{n\,\max}\left(L_{n-1}(t) - \left[L_{n-1}(t) \right]^2 \right), \qquad (95)$$

which he states eliminated the possible oscillatory behavior of successive iterations and resulted in faster convergence.

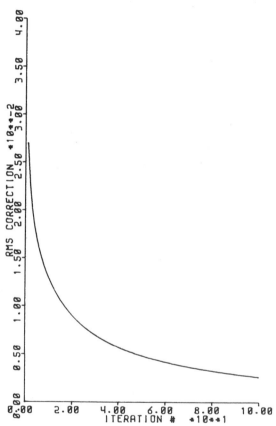

FIG. 17 Root mean square correction as a function of iteration number; a sequence of 100 iterations with $\alpha_{max} = 2.0$.

We have used the function

$$\alpha_n\big(L_{n-1}(t)\big) = \alpha_{n\,\max}\big[\,L_{n-1}(t)\big(1 - L_{n-1}(t)\big)\big]^n \tag{96}$$

most often with $n = 1$ [a factor of 4 different from Eq. (95)], but also with $n = \frac{1}{2}$, for example.

The $\alpha_n(L_{n-1})$ functions in Eqs. (94–96) all assume that the data are scaled such that maximum physically meaningful $S(t)$ is 1.00 and the minimum is 0.00. An $\alpha_n(L_{n-1})$ function that allows one to set minimum and maximum allowed values of $L_n(t)$ is given by Frieden [12] as

$$\alpha_n\big(L_{n-1}(t)\big) = c\Big[1 - 2(B - A)^{-1}\big|L_{n-1}(t) - \tfrac{1}{2}(A + B)\big|\Big], \tag{97}$$

where A is the minimum value and B is the maximum value. One can readily see that the possible relaxation parameters are limited only by one's imagination and inventiveness. A word of caution, however: The effects of a particular $\alpha_n(L_{n-1})$ must be carefully explored, since $\alpha_n(L_{n-1})$ can readily affect the area under the desired data "curve," i.e., the effect of the particular choice of $\alpha_n(L_{n-1})$ on

$$\int_t^{t+\Delta t} L_n(t)\,dt$$

is not readily predictable.

CHAPTER 7

DECONVOLUTION
EXAMPLES

Using the program DECO (cf. Appendixes 1 and 2), it is possible to illustrate the application of the deconvolution algorithm using computer-generated data. Three examples are presented. The first example uses two Gaussian lines with FWHM set to 9 units (arbitrary) separated by 18 units as shown in Fig. 18a. The input data are presented in Table II. These two lines are then convolved with a Gaussian response function (instrument function) with a FWHM of 22 units as shown in Fig. 18b. Figure 18c presents the power spectrum of the data in Fig. 18a, and Fig. 18d displays the phase spectrum of the Fourier transform of Fig. 18a. Figures 18e and 18f illustrate the corresponding Fourier transform of Fig. 18b. Notice the attenuation of the high frequencies in the power spectrum shown in Fig. 18e as compared to Fig. 18c, the effect of convolution of the desired data (cf. Chapter 2) with the response function of the measuring instrument. Figures 18g–18o display the results after 25 iterations of the algorithm discussed in Chapter 6 with $\alpha_{n\,max} = 2.0$ in Eq. (96). The error spectrum in Fig. 18i is just the algebraic difference between Fig. 18g and 18h. Note in particular the beginning of the enhancement of the previously attenuated high frequencies in Fig. 18l. Figures 18p–18x

49

7. DECONVOLUTION EXAMPLES

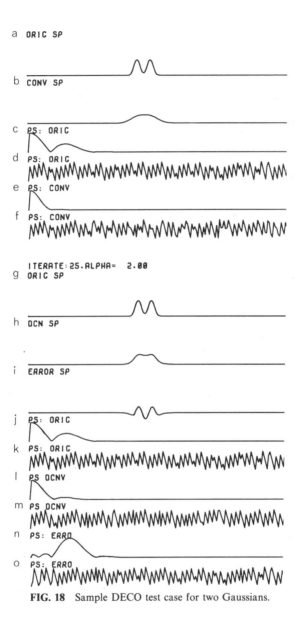

FIG. 18 Sample DECO test case for two Gaussians.

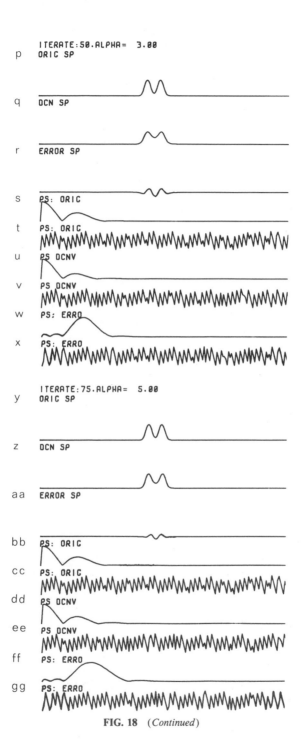

p ITERATE:50.ALPHA= 3.00
ORIC SP

q DCN SP

r ERROR SP

s PS: ORIC

t PS: ORIC

u PS DCNV

v PS DCNV

w PS: ERRO

x PS: ERRO

y ITERATE:75.ALPHA= 5.00
ORIC SP

z DCN SP

aa ERROR SP

bb PS: ORIC

cc PS: ORIC

dd PS DCNV

ee PS DCNV

ff PS: ERRO

gg PS: ERRO

FIG. 18 (*Continued*)

TABLE II

Input Data for DECO Example Case

Line	#5	#10	#15	#20	Comments
A	−2				Read 2 positions and intensities
B	150	5.			Position, Intensity, line #1
C	168	5.			Position, Intensity, line #2
D	1	9.			Line type = Gaussian, width is 9 points
E	1				Convert to absorption
F	1	22.			Response function = Gaussian; 22 points wide
G	0				No noise
H	0				No smoothing
I	−4	1.			Power spectrum, original data
J	−8	1.			Power spectrum, convolved data
K	25	2.	1	1	Do 25 iterations, set $\alpha = 2$, interpolate every other data point, use $\alpha = \alpha_0(y - 1)y$ as weight parameter
L	−3				Plot original data
M	−1				Plot current result
N	−5				Plot difference between original and current
O	−4	1.			Power spectrum, original
P	−2	1.			Power spectrum, convolved
Q	−6	1.			Power spectrum, difference

Repeat $K \rightarrow Q$ with K replaced by

	50	3.	1	1	

and then again by

	75	5.	1	1	
R	−10				Terminate

display results after 50 further iterations with $\alpha_{n\,max} = 3.0$, and Figs. 18y–18gg show the results after 75 further iterations with $\alpha_{n\,max} = 5.0$.

This example illustrates the power of deconvolution in resolution enhancement for the important yet simple case of an unresolved doublet of equal-intensity lines. Notice especially that the "spectrum" in Fig. 18b shows no structure at all and that nearly complete recovery of the doublet is achieved in Fig. 18z. The power spectra plotted by DECO are scaled to 1 in. full scale so that the power spectra illustrate frequency content accurately but the amplitudes are not directly comparable. In addition, in the phase spectrum plots, the lines that run down and to the right are pen retrace lines and not part of the phase spectrum.

The second example more closely represents a typical spectrum. The input data are described in Table III and the results are presented in Fig. 19. The case comprises a set of 35 randomly distributed lines with a FWHM = 3 units convolved with a response function of 9 units FWHM. Note particularly the progress of the restoration of the power spectrum and the phase spectrum.

The effective full width of an isolated line would be ,9.5 points, so that the actual line width of 3 points is completely masked by the hypothetical transfer function of 9 points width. Figure 19z illustrates a resolution enhancement of about a factor of 2.8. The same figure also illustrates some of the limitations of deconvolution. Notice that the detail in the original spectrum where a number of overlapping lines of width 3 points are combined such that even a factor of 5 enhancement would not resolve the full detail of the equivalent delta function or "stick" spectrum.

The third case is based on the same distribution of 35 lines as in the second example but the lines are given a FWHM = 9 units and are subjected to a response function with a FWHM = 22 units. The results are shown in Fig. 20, and comparisons between Figs. 19 and 20 are quite instructive. In particular note the dramatic effects of the reconstruction of the power and phase spectra by the deconvolution algorithm. Compare the results shown in Fig. 20dd with Fig. 20bb and those in Fig. 20ee with Fig. 20cc. Such results tend to validate the deconvolution process as viewed in Fourier transform space.

TABLE III

Input Data for DECO Example Case

Line	Data		Comments
A	35	15.	Generate 35 random lines, intensity max = 15
B	1	3.	Linetype = Gaussian; full width at half max = 3
C	1		Convert to absorption
D	1	9.	Use Gaussian response function, FWHM = 9.
E	0		No noise
F *et seq.* same as Table II, H through R			

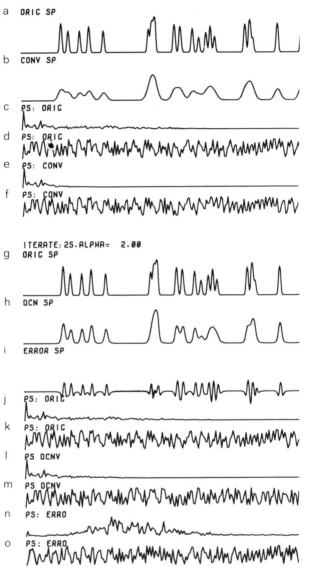

FIG. 19 Sample DECO test case output for a random spectrum containing 35 lines.

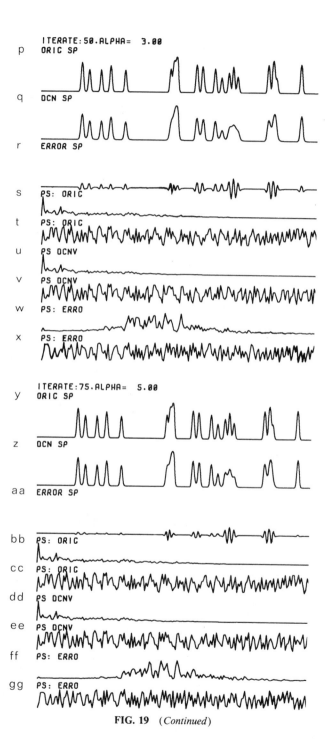

FIG. 19 (*Continued*)

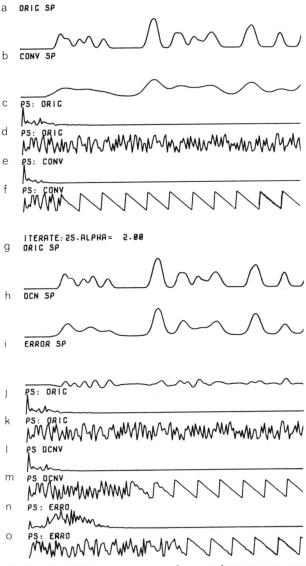

FIG. 20 Sample DECO test case output for a random spectrum containing 35 lines.

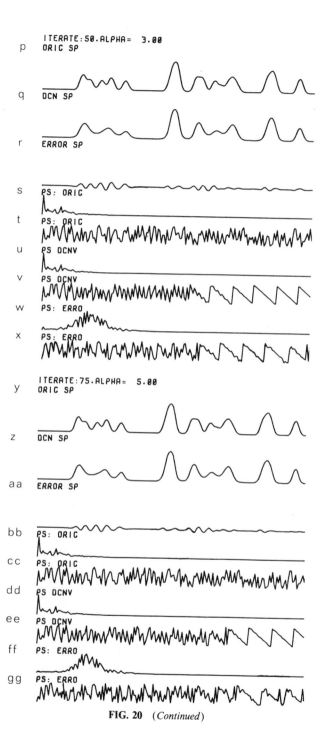

p ITERATE:50.ALPHA= 3.00
ORIC SP

q DCN SP

r ERROR SP

s PS: ORIC

t PS: ORIC

u PS DCNV

v PS DCNV

w PS: ERRO

x PS: ERRO

y ITERATE:75.ALPHA= 5.00
ORIC SP

z DCN SP

aa ERROR SP

bb PS: ORIC

cc PS: ORIC

dd PS DCNV

ee PS DCNV

ff PS: ERRO

gg PS: ERRO

FIG. 20 (*Continued*)

CHAPTER 8

TESTS AND EXAMPLES
OF DECONVOLUTION

The original goal of our study of deconvolution techniques was to test the validity and reliability of the results of deconvolution of high-quality infrared absorption spectra. An early concern involved the positional stability of transitions under deconvolution. Since molecular spectroscopy studies rely on accurate transition frequencies in order to fit the molecular model to the spectrum and thereby extract parameters characteristic of the molecule, deconvolution procedures are useful only if transition frequencies are stable under deconvolution.

The general aim of this first series of tests carried out in 1973–1974 was to record high-quality spectral records at various slit widths (and thereby various resolutions) and compare the deconvolved results from the lower-resolution data to data recorded at higher resolution. The spectra used in this series of tests is a section of the Q branch of ν_2 of CHD_3 from 2133 cm^{-1} to 2144 cm^{-1} [23]. Data were recorded at slit widths of 1000 μm, 600 μm, and 300 μm with several scans at each slit width averaged to improve the signal-to-noise ratio. A single scan at 150-μm slits was recorded as the experimental benchmark for the frequencies.

59

The response function was determined after Jansson [24] using P(16) of the 1–0 band of CO at 2082 cm^{-1} recorded for each of the three slit widths at the same scanning parameters used for the spectral data.

About 170 lines were selected from the 150-μm slit width run and the positions of each of these lines in the three deconvoluted spectra were measured and compared to the measured frequencies from the 150-μm run. The standard deviation of the frequency variations for the 65 best transitions compared to the 150-μm run was 0.002 cm^{-1} or less and the standard deviation of 510 lines from three deconvolved spectra was 0.0037 cm^{-1}.

Tests made at the same time indicated that the capability to fit one scan to another had a comparable standard deviation, i.e., approximately 0.002 cm^{-1}. Figure 21 shows a section of the spectra used for the test; Table IV presents the experimental parameters.

Since these tests were carried out, we have improved the optomechanical system and now are able to fit successive 150 cm^{-1} long data records to one another with a standard deviation of 0.8 to 0.5×10^{-3} cm^{-1}. Additional line position stability tests carried out routinely on working spectral records indicate positional (wavenumber) stability under deconvolution of better than 0.5×10^{-3} cm^{-1}.

In a recent report, Pliva *et al.* [13] have carried out independent tests of deconvolution using the Jansson algorithm and have reached similar conclusions regarding positional stability of spectral features.

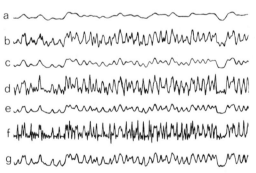

FIG. 21 Positional stability test spectra. Experimental data at resolution of 0.055, 0.033, 0.020 and 0.011 cm^{-1} in a, c, e, g, respectively and deconvolved data immediately below the experimental data in b, d, and f.

TABLE IV

Line Position Stability Test: Scan Parameters

Slit width (μm)	Spectral slit width (cm^{-1})	Number of scans averaged	Scan rate (cm^{-1}/day)	Electronic band width (Hz)
1000	0.055	3	36	0.125
600	0.033	5	36	0.042
300	0.020	7	36	0.013
150	0.011	1	18	0.004

Other questions concerning the stability of deconvolution proce-dures against generation of spurious spectral features were studied. In general, deconvolution properly carried out is quite stable. It is, however, possible to generate spurious features if a response function significantly different from the actual response function is used. In addition, attempts to deconvolve a noisy spectral record can generate significant spurious features. Reference to Fig. 15c can illuminate this difficulty. When the power spectral density of attenuated spectral frequencies (above f_n in the transform space of Fig. 15c) lies below the power spectral density of the noise, the information relating to these higher spectral frequencies is severely corrupted, and attempts to enhance these frequencies will result in unpredictable results—generally in the form of spurious spectral features in signal space.

Our experience with Doppler-limited spectra of $^{12}CD_3F$ in the 5-μm region of the spectrum verifies the results of the noise sensitivity tests. In addition, even when *weak* spurious noise-generated features are produced because of excessive deconvolution relative to the recorded signal-to-noise ratio, the validity of the stronger spectral features is not lost, and in fact the positional stability of the true spectral features is not markedly affected. These results appear to hold when noise-generated features have fractional absorption of 0.1 or less on a scale of 0.0–1.0 and where spectral features used in analysis have a minimum fractional absorption of 0.25.

Figure 22a shows a section of the Q branch of $^{74}GeH_4$ obtained in our laboratory. Figure 22b shows the experimental data for two different resolution values in the first and third trace and the decon-volution results in the second and fourth traces. Comparison of the deconvolved 0.020 cm^{-1} trace (at a resolution of 0.008 cm^{-1}) with the actual 0.010 cm^{-1} trace lends validity to the preceding statements. In

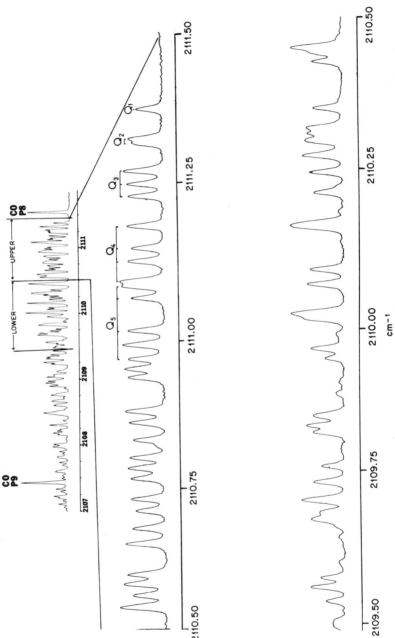

FIG. 22a Section of the Q branch of $^{74}GeH_4$ obtained at the University of Tennessee. Some assignments are indicated.

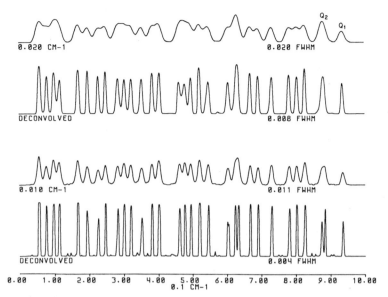

FIG. 22b Two experimental records, first and third corresponding to a section of that shown in Fig. 22a. The second and fourth traces are deconvolved traces of the first and third, respectively.

all test cases run in our laboratory with appropriate response functions, no spurious transitions have been generated.

Finally, in Fig. 23 we present a section of ν_4 of CD_3F with a section shown at 0.010 cm^{-1} (10 mK) resolution equivalent to 4 N slits [25], deconvolved to a measured FWHM (full width at half maximum) of 0.0045 cm^{-1} (4.5 mK), which is 1.83 times better apparent resolution than that achievable at infinitely narrow or zero N slits. Validation of this "superresolution" result is difficult in an absolute sense. However, a 0.004 cm^{-1} (4 mK) calculated spectrum is shown based on a preliminary analysis of the nondeconvolved full band data. It is interesting to note that the Doppler width of the spectral lines of CD_3F in this region is 0.0046 cm^{-1}. Thus, deconvolution has allowed us to obtain Doppler-limited spectra of CD_3F even though zero N slit spectra would yield spectra of an apparent resolution of 0.0082 cm^{-1} ($[0.0068^2 + 0.00462^2]^{1/2}$).

Recent tests by Pliva *et al.* [13] using a Doppler-limited spectrum of C_6H_6 measured in a difference-frequency laser system resulted in deconvolved line widths of $\sim 1.2 \times 10^{-3}$ cm^{-1} (compared to the

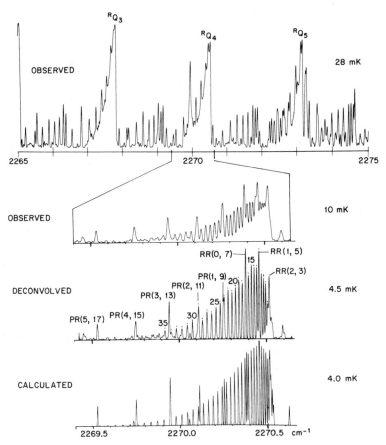

FIG. 23   Section of the spectrum of CD_3F obtained at the University of Tennessee showing actual resolutions of 0.028 cm^{-1} (28 mK), 0.010 cm^{-1} (10 mK) and a trace of the 10-mK data deconvolved to 4.5 mK.

Doppler width of 3.6×10^{-3} cm^{-1} at 203°K). The reliability of their procedures was verified by extensive testing that indicated that at signal-to-noise ratios on the order of 1000:1, resolution enhancements of a factor of 3–5 were possible.

Deconvolution, viewed as discussed in the introductory chapters, can quite readily be used to achieve such "superresolution" results. However, much care must be exercised in order to avoid generation of nonphysical results.

CHAPTER 9

NOISE

In Chapters 2 and 3 we discussed the interactions of the physical measurement system with the physical system under investigation. It was pointed out that noise would always be present and would be the primary complication limiting the success of deconvolution. In this chapter we discuss noise and precisely how and why it affects deconvolution. We also investigate experimentally and theoretically the effects of noise on the resolution enhancement that can be achieved.

Noise can be defined as any unwanted signal that somehow gets added to the system output signal. In the specific example of Chapter 3, noise actually creeps into the picture at every stage of the measurement system, starting with the "white light" source the output of which may fluctuate in time to the detector, which adds its contribution to the noise from random electron motions and fluctuations in the background radiation falling on the detector. The signal amplification system adds electronic noise that becomes more apparent at the higher gain settings necessary when very narrow slits are used. These problems can be reduced in a variety of ways, but they cannot be eliminated. There will always be noise.

In infrared spectroscopy the primary source of noise is the detector noise. Detector noise is said to be "white noise," that is, it contains all frequencies. If we examined the Fourier transform of the detector noise signal, we would see that all frequencies are present in the power spectral density. Detector noise is greatly reduced by using cooled detectors and by chopping the light beam at a fixed frequency, which makes it possible for a lock-in amplifier to reject most of the noise. The electronics further reduces the noise in several ways. First of all, noise frequencies that are very high are cut off by the amplifier and electronic filters. Noise of intermediate frequencies, which must be passed by the electronics if the true signal is not distorted, present the experimenter with unavoidable problems. One way around this is to access data points more slowly. This pushes the spectral power representing the real signal to lower frequencies so that the signal will not be distorted if the problematic intermediate frequencies are attenuated by the electronic filtering system.

There is always a frequency region, however, where the noise signal cannot be eliminated without affecting the true signal. This is of particular concern to someone interested in deconvolution, since it is in this frequency range that we need to enhance the signal in order to improve the resolution. (Recall the examples of Chapter 7.) Noise is the primary reason that transform or frequency space deconvolution algorithms do not work well and is generally the limiting factor in signal (i.e., time) space deconvolution.

It follows that the problems of noise are complex in general and for deconvolution, difficult to attack on a theoretical basis. We will proceed in a more direct fashion. By generating a spectrum (a hypothetical system output signal) to deconvolve, we know what the spectrum should look like if the deconvolution is successful. We can also generate a hypothetical noise signal so that the signal-to-noise ratio is known. Even in this highly controlled case, it is still difficult to make quantitative statements on the effects of noise. Except in the case of extreme noise, the algorithm described by Jansson will not diverge. However, it is apparent that improving the signal-to-noise ratio improves the deconvolution results. Shown in Fig. 24 are several cases where identical noise-free signals were corrupted by various amounts of noise before deconvolution. Figure 25 displays the corresponding root-mean-square correction as a function of iteration number for each case. By use of the program outlined in Appendix 1, it is hoped that researchers interested in deconvolution can determine the noise requirements necessary for their specific measurement system and their specific goals. From these tests, as well as from our experience in deconvolving actual spectral data, it is apparent that a

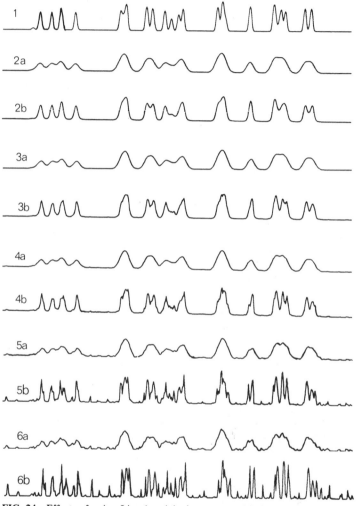

FIG. 24 Effects of noise. Line 1: original spectrum with Gaussian line width of 3. Lines 2a, 3a, 4a, 5a, and 6a: The original spectrum convolved with a Gaussian of width 10 and noise added to produce $S/N \approx \infty$, 100, 50, 20, and 10, respectively. Lines 2b, 3b, 4b, 5b, and 6b: The deconvolved results for 50 iterations with $\alpha_{max} = 1.5$.

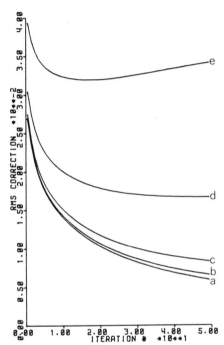

FIG. 25 Root-mean-square correction for noisy data: a, b, c, d, and e correspond to 2, 3, 4, 5, and 6 of Figure 24.

signal-to-noise ratio of approximately 100:1 is required to achieve a deconvolution by a factor of 2–2.5. Pliva *et al.* [13] find that a S/N ratio of 1000:1 is required for a factor of 3.5–5. The type of results that one can expect with signal-to-noise ratios less than 100:1 are shown in Fig. 24. Note that Fig. 24 does not purport to show the potential of deconvolution, since only a limited deconvolution was carried out, but rather to show the sensitivity of the technique to spurious feature generation due to the presence of noise.

It should be further noted that the signal-to-noise ratio in the case of absorption spectroscopy is the ratio of the 100% transmission (0% absorption) signal to the noise signal. Thus, it is the actual noise amplitude that poses a problem and not the noise amplitude relative to the amplitude of spectral features that might well be called the information-to-noise ratio. In actuality, this is a fortunate result for deconvolution users, since the signal-to-noise ratio is normally much greater than the information-to-noise ratio mentioned earlier.

A natural question to ask when considering the signal-to-noise ratio relates to filtering or smoothing of the spectral data record. One might ask why not simply smooth the data record before deconvolution; in fact, why not smooth or filter the record until a satisfactory signal-to-noise ratio is achieved and then proceed with deconvolution.

The answer to this query is twofold. First, it is appropriate to smooth or filter the spectral data record before deconvolution, as we shall discuss. Second, there is a practical limit to the amount of smoothing that makes sense.

For purposes of this discussion of noise and smoothing, let us agree that the spectrum we wish to deconvolve shall be called the spectral data record—a time series in signal space. Let us further agree that spectral frequencies shall refer to the Fourier-transformed power spectrum in frequency space. Spectral features shall refer to the details of the signal space spectral data record; spectral frequencies shall refer to the power spectrum in transform or frequency space.

Smoothing and filtering act on the power spectrum in such a way as to attenuate the spectral power density at certain frequencies. If high frequencies are attenuated, it is called low pass filtering, which is the type normally employed in smoothing a spectrum. The attenuated high frequencies in the power spectrum correspond to sharp, high-resolution features in signal space. If we act on the spectral data record in such a way as to attenuate high spectral frequencies, the spectral data record will have fewer sharp features. Thus, as we

attenuate the noise in the spectral record we may also significantly
degrade the resolution of features in spectral record.

In addition, deconvolution—as discussed in earlier chapters—is a
process that restores previously attenuated high frequencies in the
power spectrum of the spectral data record. Thus, deconvolution and
smoothing work at cross purposes.

Smoothing works to our benefit only as long as the frequencies
attenuated do not significantly contribute to the spectral features we
wish to enhance. This places a practical limit on the signal-to-noise
enhancement that is achievable by filtering and smoothing. As was
mentioned earlier in this chapter, the acquisition rate of the spectral
data record may be reduced, thus moving the spectral frequency
content of the power spectrum of the spectral data record to lower
frequencies and away from damaging high-frequency noise power.

Smoothing of the spectral data record has been treated by Willson
and Edwards [22] in an extensive set of tests of the Savitsky and
Golay least squares smoothing technique [26] using transversal filters
[15].

In our laboratory, we use a two-stage digital RC filter in addition
to the 12-dB roll-off RC filter incorporated in the lock-in amplifier.
This filter gives us somewhat more precise control over the effective
bandwidth of our amplifier/analog to digital conversion system.

A single-stage low pass filter may be specified by a first-order
difference equation of the form [1, 15]

$$R^+(k) = (1 - \alpha)R(k) + \alpha R^+(k - 1), \qquad (98)$$

where $R(k)$ is the kth spectral sample (digitized output of the lock-in
amplifier) and $R^+(k)$ is the kth filtered sample. For a specified band
width ω_1 and sampling period T (sec), α is given by

$$\alpha = 2 - \cos\omega_1 T - \left[(3 - \cos\omega_1 T)(1 - \cos\omega_1 T)\right]^{1/2}. \qquad (99)$$

Bandwidth is defined by the half-power point, that is, a sinusoid of
frequency ω will be passed by the filter at half power. (The signal
power of $A \sin kT$ is given by $A^2/2$.) The gain factor is provided by
the filter transfer function [15]

$$H(z) = \frac{(1 - \alpha)z}{z - \alpha} \qquad (100)$$

or specifically by

$$H\left[\exp(j\omega t)\right] = \frac{1 - \alpha}{\left(1 + \alpha^2 - 2\cos\omega T\right)^{1/2}}. \quad (101)$$

The gain factor describes the effect of the filter on any frequency. Stable filters are those for which $\alpha < 1$.

Superior high-frequency roll-off may be obtained with n cascaded filters. The bandwidth determines the value of α according to

$$\alpha = \frac{1}{2^{1/n} - 1}\left\{2^{1/n} - \cos\omega_1 T - \left[\left(2^{1/n} - \cos\omega_1 T\right)^2 - \left(2^{1/n} - 1\right)^2\right]^{1/2}\right\}. \quad (102)$$

Filter response as a function of α for a given ω_1 is determined by the nth power of the right-hand side of Eq. (101). The filter is given by [15]

$$R^+(k) = \sum_{i=1}^{n} (-1)^{i+1}\alpha^i\binom{n}{i}R^+(k - i) + (1 - \alpha)^n R(k). \quad (103)$$

Before further processing of the spectral record, a transversal filter of the Savitsky–Golay type is applied to the record. We normally use a filter length of one-half of a line width in data points. If additional smoothing is indicated we apply the same filter a second time. The Savitsky–Golay filter is useful for high-frequency noise reduction but cannot be used to remove spectral line–like features with "widths" approximating real spectral lines without seriously degrading the spectral data record.

CHAPTER 10

COMPLETE CASE HISTORY OF A DECONVOLVED DATA RUN

In order to tie all the preceding together and fill in any missing details, this chapter contains the complete case history of a data run that has been deconvolved and analyzed. The data set to be described is a high-resolution spectrum of $^{12}CD_3F$ in the 5-μm spectral region. The spectrum of ν_1 was run at a resolution of 17×10^{-3} cm^{-1} (17 mk) from 2060 cm^{-1} to 2140 cm^{-1}. The spectrometer was the 5-m focal length Littrow spectrometer at the University of Tennessee [27]. The spectrometer is a grating instrument capable of a resolution of 0.0069 cm^{-1} at 65° incidence on a 20×40 cm diffraction grating ruled with 31.5 rulings/mm. At slit widths of 4 N [25] the instrument has its highest practical resolution of 0.0097 cm^{-1} at 65° using an Abbe criterion [25].

The data set to be described covered the region from 2060 cm^{-1} to 2140 cm^{-1}. At a resolution of 0.009 cm^{-1} the run would have taken 595 hr. Since the Doppler width of the spectral lines is 0.004

cm^{-1}, it was decided to run the spectrum at a resolution of 0.018 cm^{-1} and to deconvolve to 0.0065 cm^{-1}, nearly the Doppler width. The actual data run was made at a scanning rate of 4.3 cm^{-1}/hr for a total of 18.6 hr per scan. The data, for various reasons, were run three times and the resulting scans averaged point-by-point. Because we have a stepping motor drive on the tangent arm of the grating [1], it is possible to average successive scans to improve reliability and of course the signal-to-noise ratio. The amplifier time constant was 1.25 sec, making the amplifier passband 0.1 Hz. Each data point was sampled 2400 times, which resulted in an acquisition rate of approximately 2 sec per data point. The entire scan comprises 42,650 data points. These parameters are consistent with a sampling frequency of 2.4π times the minimum Nyquist frequency required to pass 99.5% of the spectral power density in a 0.018 cm^{-1} FWHM spectral line [14]. The sample path length was 1 m and the $^{12}CD_3F$ pressure was 500×10^{-3} torr. A calibration gas, CO, was also included in the absorption cell at a pressure of 50×10^{-3} torr.

The actual data sampling was accomplished by averaging 32 samples per observation and by making 75 observations at each data point. The two-stage digital RC filter resulted in mild increased filtering with a passband of approximately four times that of the amplifier. The digital filtering serves to further reduce high-frequency noise content in the digitally recorded spectrum. The signal-to-noise ratio for each of the three spectra was approximately 50:1 to 75:1. The data were recorded in a FORTRAN readable file of 250 words per record. The raw data record is displayed in Fig. 26.

Each of the three records was smoothed using a Savitsky–Golay [26] transversal filter with a width of 5 points. The process was carried out twice for each of three records.

At this point it was necessary to remove any artificial constant signal from the record—i.e., to remove the baseline. For our data this is necessary because 0% absorption corresponds to a digital signal value of 32,768. This must be converted to zero and any stray light signal (of which we have none detectable) must also be removed. If any constant amplitude signal remains, deconvolution will produce side lobes on each spectral line. We use a simple baseline removal procedure; we use a point plot display and a light pen to measure data point number and baseline amplitude at a number of points in the spectrum and then generate a piecewise straightline baseline curve, subtract it out, and rescale the resulting data to a range of 0 to

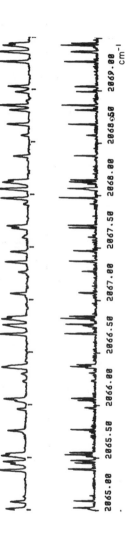

2065.00 2065.50 2066.00 2066.50 2067.00 2067.50 2068.00 2068.50 2069.00
cm^{-1}

FIG. 26 Case history spectral display; high-resolution infrared spectrum of $^{12}CD_3F$, ν_1 band. Top trace is raw data as recorded by spectrometer control computer. Second trace from top is the average of three independent data runs, averaged after the baseline was removed from each data run. The bottom trace is the deconvolved spectrum exhibiting a resolution enhancement of approximately a factor of 3. The resolution in the deconvolved trace is about 0.005 cm^{-1} to 0.006 cm^{-1}, whereas the resolution in the second trace from the top is approximately 0.017 cm^{-1}. Resolution is measured as full width at half maximum for the narrowest transitions.

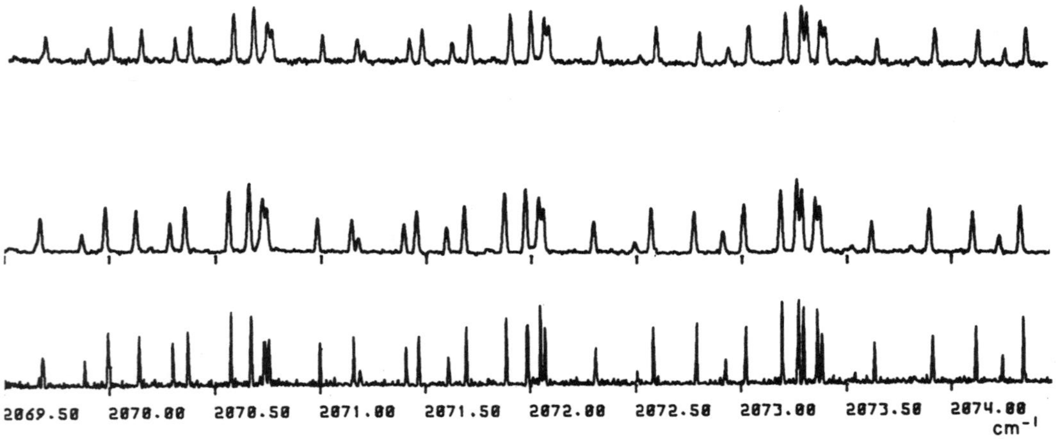

FIG. 26 (Continued)

2069.50 2070.00 2070.50 2071.00 2071.50 2072.00 2072.50 2073.00 2073.50 2074.00
cm⁻¹

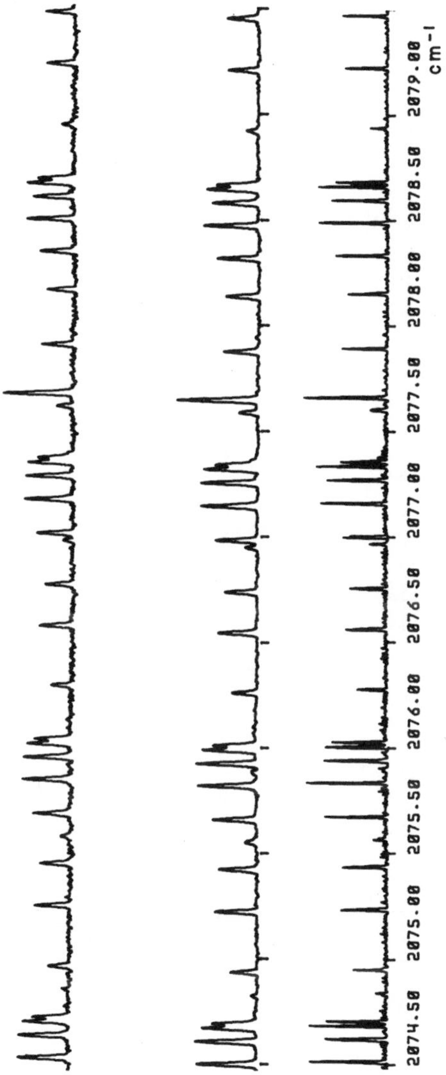

FIG. 26 (Continued)

2074.50 2075.00 2075.50 2076.00 2076.50 2077.00 2077.50 2078.00 2078.50 2079.00
cm⁻¹

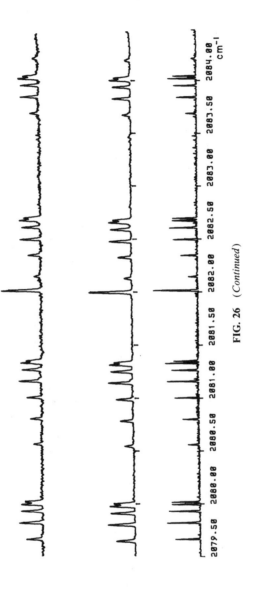

2079.50 2080.00 2080.50 2081.00 2081.50 2082.00 2082.50 2083.00 2083.50 2084.00
cm⁻¹

FIG. 26 (Continued)

FIG. 26 (*Continued*)

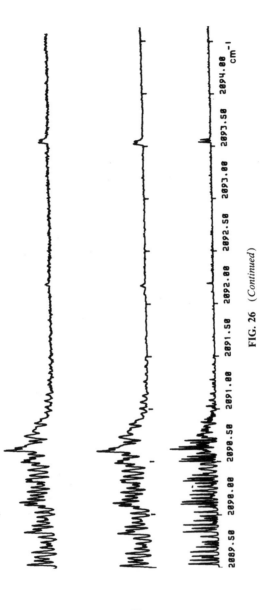

FIG. 26 (*Continued*)

82

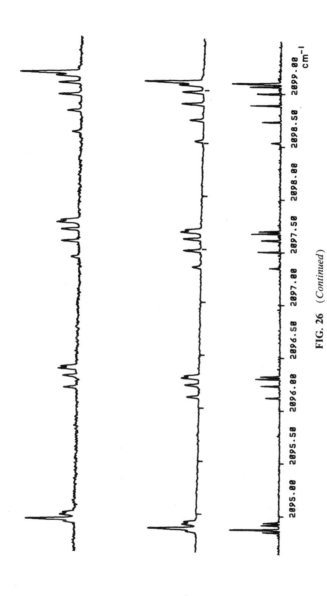

2095.00 2095.50 2096.00 2096.50 2097.00 2097.50 2098.00 2098.50 2099.00 cm^{-1}

FIG. 26 (Continued)

83

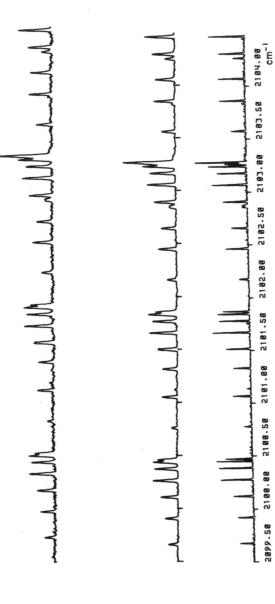

FIG. 26 (Continued)

84

FIG. 26 (*Continued*)

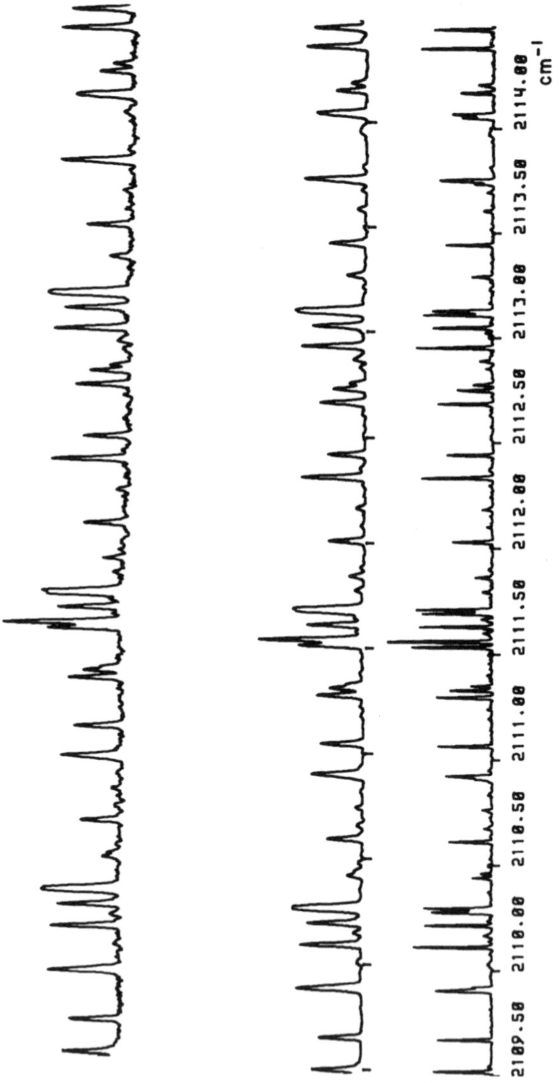

FIG. 26 (Continued)

2109.50 2110.00 2110.50 2111.00 2111.50 2112.00 2112.50 2113.00 2113.50 2114.00 cm⁻¹

86

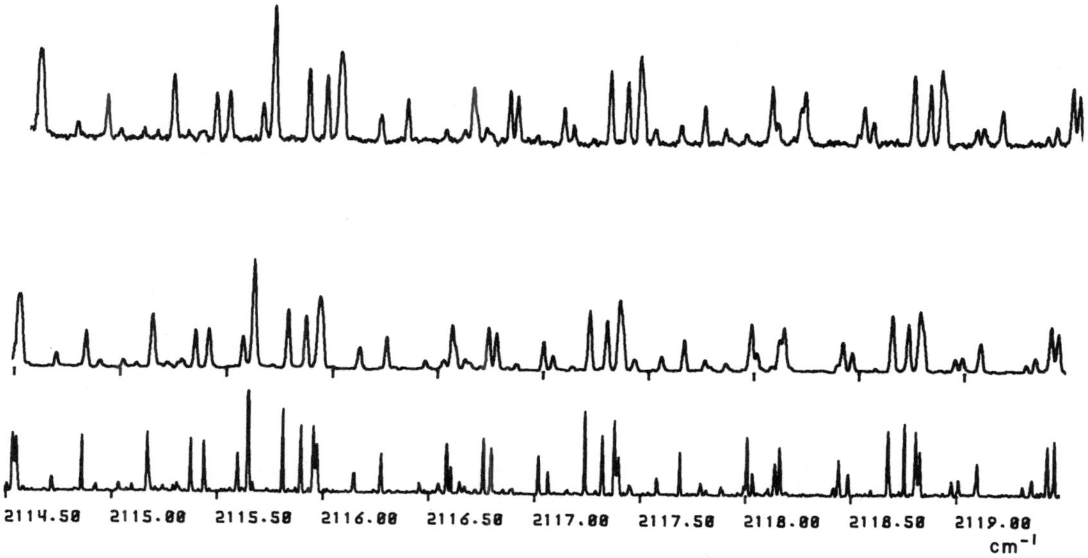

FIG. 26 (Continued)

2114.50 2115.00 2115.50 2116.00 2116.50 2117.00 2117.50 2118.00 2118.50 2119.00
cm⁻¹

87

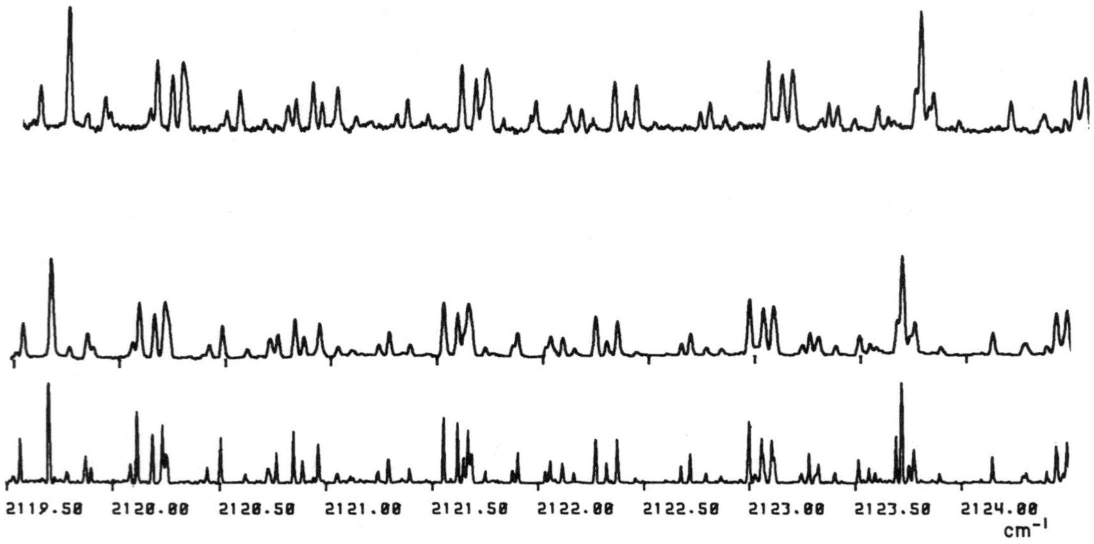

FIG. 26 (*Continued*)

2119.50 2120.00 2120.50 2121.00 2121.50 2122.00 2122.50 2123.00 2123.50 2124.00
cm^{-1}

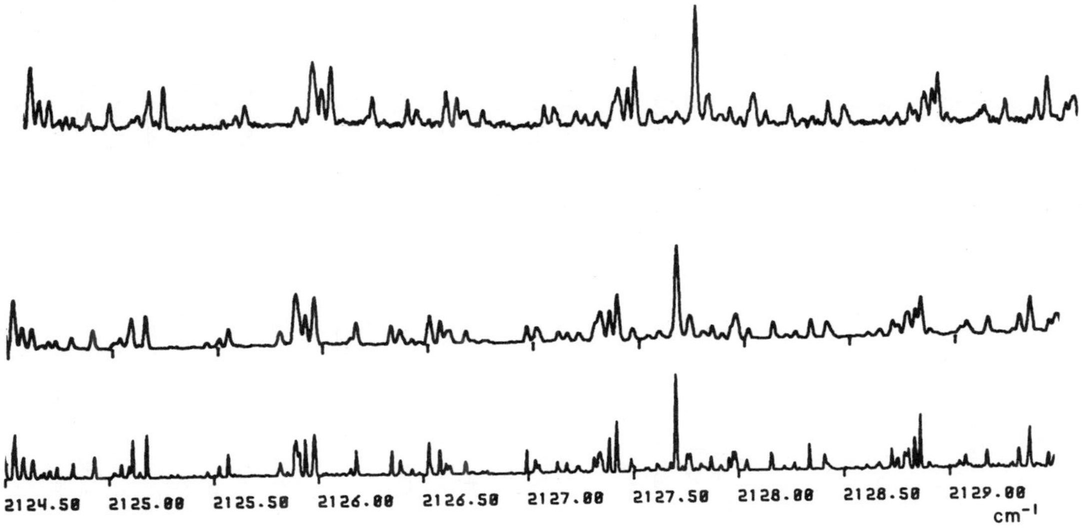

FIG. 26 (Continued)

2124.50 2125.00 2125.50 2126.00 2126.50 2127.00 2127.50 2128.00 2128.50 2129.00 cm⁻¹

89

32,768. Several other baseline correction procedures have been tried, including fitting a polynomial to a set of baseline points. The point-by-point method seems to work best for our spectra. We have given some thought to the use of a cubic spline [28] for baseline removal as well as a low-frequency filtering method developed by Atakan, Blass, and Jennings [29] for diode laser spectra. These latter techniques have not been tried for our laboratory spectra as yet.

The three runs were then averaged on a point-by-point basis and the averaged spectrum is shown in Fig. 26. (We do not always average spectra as a routine matter, although we did in this case.) The resulting signal-to-noise ratio in the averaged spectrum is approximately 150:1.

For long file deconvolution we use the University of Tennessee Computing Center PDP 10 system with a KL processor in batch mode. The program used is LONGO, described and listed in Appendix 4. The only changes are the required file specification statements. The PDP-11 file specification statements CALL ASSIGN are replaced with OPEN statements in PDP-10 FORTRAN.

We transport the spectra from PDP-11 to PDP-10 via DECTAPE. The unformatted RT-11 (operating system) data files are converted to formatted FORTRAN files for transport to the PDP-10. Once stored in the PDP-10, the formatted files are reconverted to unformatted files for processing. As these FORTRAN utility programs are machine and local option dependent, they are not included in the appendixes. (Listings are available from the authors.)

A separate program that runs on the PDP-10 produces a macro interpreted command (MIC) file for the batch processor version of LONGO (cf. Appendix 4). The program is interactive and used in the time-sharing mode and allows the user to answer appropriate questions concerning the deconvolution run.

We are thus at the point where the averaged data file is to be deconvolved. We chose to use the following sequence of iterations: $\alpha = 0.5$, 5 iterations; $\alpha = 1.0$, 10 iterations; $\alpha = 2.0$, 20 iterations; $\alpha = 4.0$, 40 iterations. We have found that for our data with a signal-to-noise ratio exceeding 100:1, these parameters produce a resolution enhancement of 2:1 to 3:1 without significant noise enhancement.

For this deconvolution run, the root mean square error, RMSE, (cf. Chapter 6) started at a value of 471 (full scale being 32,768) and reached values of 418, 296, 161, and 80 at the end of each different α cycle.

The deconvolved spectrum is shown in Fig. 26. The resolution is 0.006 cm^{-1}, an enhancement of a factor of 3.

The deconvolved data set is converted to a formatted data set for transport to our PDP-11 system, where it is measured, calibrated, plotted, and analyzed in-house.

Running on the PDP-10, KL processor at the University of Tennessee, the deconvolution of the averaged data was accomplished in 22.50 min using a response function of 9 points FWHM, 36 points wide total. The raw spectrum exhibited a width of 9–10 data points FWHM (1 point ≈ 0.002 cm^{-1}). We have found that the deconvolution time on the KL processor is given by approximately 2×10^{-7} min per spectrum point per response function full width ($\approx$ four times response function FWHM) per iteration. The deconvoluted spectrum is 42,650 points long, the response function width is 36 (4 times 9 points FWHM), the number of iterations is 75. Thus, the estimated total time is given by $[42{,}650 \cdot 36 \cdot 75 \cdot 2 \times 10^{-7}] = 23.03$ min.

CHAPTER 11

DECONVOLUTION AND INTENSITIES

When deconvolving spectra, the question of the validity of deconvolved line intensities naturally arises. This is true whether or not one intends to use deconvolution techniques for definitive intensity studies.

Several studies have been carried out using the program DECO described and listed in Appendixes 1 and 2. The studies used the weighting function α given in Eq. (96), which limits the deconvolved results to the range from 0% absorption to 100% absorption. In actuality α is entered into our programs as $\alpha = \alpha_{max}$ times a weight function, which is given by Eq. (96) divided by α_{max}.

The results for single lines of Gaussian and Lorentzian shape broadened by a Gaussian response function are given in Table V. For the Lorentzian lines, the last iteration cycle of $\alpha = 4$, 40 iterations seem to begin to distort the line shape in a nonphysical fashion. The best line shape for Lorentz lines is obtained at the end of the $\alpha = 2$ cycle. The results are reasonably encouraging, since most intensity studies have an accuracy of 2–20%. Thus, deconvolution seems to preserve the intensity (area) quite well when the response function is reasonably completely removed, that is, when the deconvolved line is nearly the width expected for a zero width response function. Even after mild deconvolution at the end of the $\alpha = 1$ cycle the accuracies are 7%, 3%, 4%, and 0.5% for cases 1–4, respectively (cf. Table V). Figures 27–30 present the DECO output for cases 1–4 of Table V. Table VI lists the DECO input data used for cases 1–4.

Attempts were made to repeat the cases in Table IV with a weighting (α function) such that α was a constant, independent of signal amplitude. As usual for such weighting, negative side lobes developed and the results were not considered valid.

It is not expected, nor is it borne out in practice, that deconvolution recovers the detailed line shape of the original lines. The results are not extremely different from the original shapes, but one should not in general expect to use deconvolved line shapes to obtain significant information about intramolecular potentials.

On the other hand, some preliminary tests on Lorentzian lines of 20 points FWHM broadened by a Gaussian response function of 10 points FWHM seem to indicate that for the practically significant case in molecular spectra, i.e., typically described immediately above, deconvolution seems to recover a good bit of the information about the true line profile. Figure 31 displays a Lorentzian line (the Lorentzian line has a width of 20 points) broadened by a Gaussian response function of width 10 points. The successive iterations are indicated in the figure as well as a display of the error spectrum that is the original spectrum minus the deconvolved approximation. Figure 31 is presented principally for shape preservation information under the influence of mild deconvolution. Cases 1–4 discussed earlier show that area is ultimately preserved for single transitions.

Finally, we present several studies of Gaussian doublets broadened by Gaussian response functions. Figures 32 and 33 both present studies of doublets made up of 8-point-wide Gaussian lines, separated by 25 points, and broadened by 16-point-wide Gaussian response functions. Notice that in both cases, the noninstrumentally broadened lines are baseline resolved and thus we can measure the area of the lines. Once these are instrumentally broadened, the area of the lines is undefined. In the equal-intensity doublet shown in Fig. 32, the original spectrum area of each line is 8.50, each has a width of 8 points. After the $\alpha = 4.0$ iteration, the area of each is 8.55 and the width is 8.1 points. The accuracy for area measurements in this case is 0.6%.

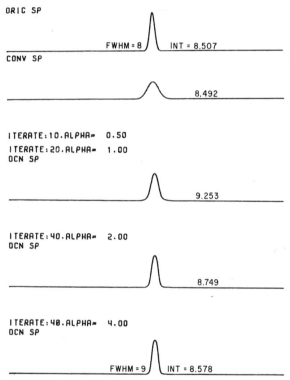

FIG. 27 DECO output for intensity test, case 1; line is Gaussian, 8 points FWHM; response function is Gaussian, 16 points FWHM.

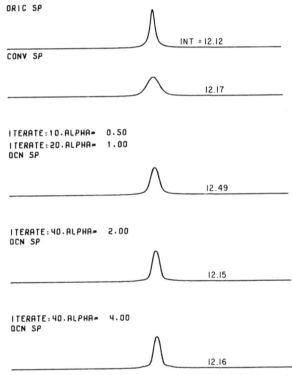

FIG. 28 DECO output for intensity test, case 2; line is Lorentzian, 8 points FWHM; response function is Gaussian, 16 points FWHM.

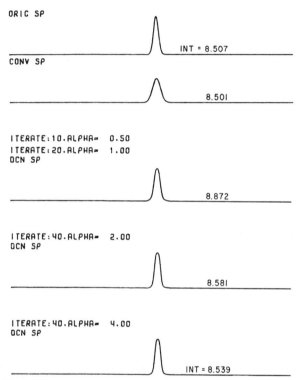

FIG. 29 DECO output for intensity test, case 3; line is Gaussian, 8 points FWHM; response function is Gaussian, 10 points FWHM.

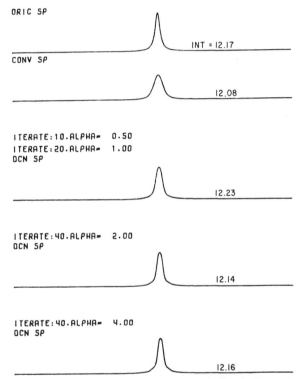

FIG. 30 DECO output for intensity test, case 4; line is Lorentzian, 8 points FWHM; response function is Gaussian, 10 points FWHM.

TABLE V

Intensity Deconvolution Tests Using DECO for Four Cases[a]

Case	1	2	3	4
Line width (FWHM)	8	8	8	8
Line shape	Gaussian	Lorentz	Gaussian	Lorentz
Intensity	8.51	12.12	8.51	12.17
Response F. width	16	16	10	10
Response F. shape	Gaussian	Gaussian	Gaussian	Gaussian
"Observed" intensity	8.49	12.17	8.50	12.08
"Observed" width	18	18	12	12
Cycle 1:	0.5	0.5	0.5	0.5
Iterations	10	10	10	10
Intensity	9.21	12.67	8.97	12.40
FWHM	16	17	10	10.6
Cycle 2:	1.0	1.0	1.0	1.0
Iterations	20	20	20	20
Intensity	9.25	12.49	8.87	12.23
FWHM	12	11	9.5	10.5
Cycle 3:	2.0	2.0	2.0	2.0
Iterations	40	40	40	40
Intensity	8.75	12.15	8.58	12.14
FWHM	10	11	8.5	9.6
Cycle 4:	4.0	4.0	4.0	4.0
Iterations	40	40	40	40
Intensity	8.56	12.16	8.54	12.16
Weight type	1	1	1	1
Final width	8.7	10	8.3	8.9
Error (in intensity)	0.6%	0.3%	0.4%	0.1%

[a] Each case is a single isolated spectral line.

TABLE VI

**DECO Input Data for Single Line Intensity Studies,
Cases 1–4**

Case 1:	− 1			
	20028000			
	1	8.		
	0			
	1	16.		
	0			
	10	0.5	1	1
	− 1			
	20	1.	1	1
	− 1			
	40	2.	1	1
	− 1			
	40	4.	1	1
	0			
Case 2:	− 1			
	20028000			
	2	8.		
	0			
	1	16.		
	0			
	10	0.5	1	1
	− 1			
	20	1.	1	1
	− 1			
	40	2.	1	1
	− 1			
	40	4.	1	1
	− 1			
	0			

Case 3: Same as Case 1 except change 16. in line 5 to 10.

Case 4: Same as Case 2 except change 16. in line 5 to 10.

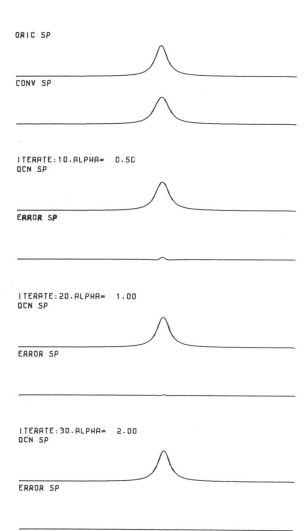

FIG. 31 Lorentzian line (20 points FWHM) broadened by a 10-point FWHM Gaussian and deconvolved. Error spectrum shows progress and partial success of shape (profile) recovery.

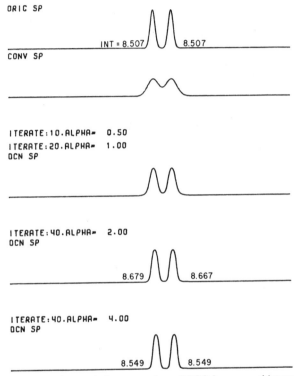

FIG. 32 Doublet deconvolution output from DECO for an equal intensity doublet. Intensity (area) is preserved after $\alpha = 4.0$ iterations to an accuracy of 0.6%.

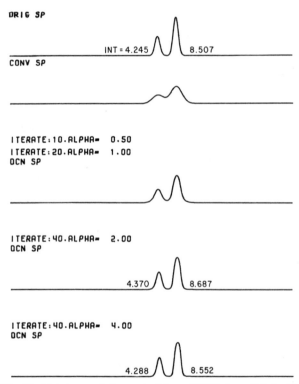

FIG. 33 Doublet deconvolution output from DECO for a 2:1 doublet. Intensity (area) is preserved after $\alpha = 4.0$ iterations to an accuracy of 0.9% (left member) and 0.6% (right member).

In Fig. 33 the original areas of the members of the doublet are 4.25 and 8.50 and after the $\alpha = 4.0$ iterations, the areas are 4.29 and 8.55 for accuracies of 0.9% and 0.6%, respectively.

In general, our tests seem to indicate that deconvolution is usable for intensity studies, if the signal-to-noise ratio is sufficiently high so that the response function may be nearly completely removed from the data. This limits us to resolution enhancement cases of about a factor of 3 in most experimental situations.

When only mild deconvolution is required to remove the effect of the response function, shape information seems to be partially recovered, although the shape tests are far from definitive at this time.

CHAPTER 12

ADVICE

Researchers beginning to use deconvolution techniques may well encounter difficulties that have become familiar to us. A summary of our experiences in this area may provide new users with additional insights that should help them avoid common pitfalls. It may also help users to identify and correct several more subtle difficulties that do arise.

Many of the problems are similar in character to the noise problems discussed in Chapter 9. It is difficult to predict the exact character of many of these problems. This is due to the effect of variations in data sets to be deconvolved and also due to the fact that no single set of iterative parameters can be specified for use in all cases. We can say that when these difficulties do arise, the deconvolution algorithm will not work as well as it should. In the extreme case of inadvertant neglect of potential problems, the algorithm may diverge or produce nonphysical results. In the latter instances it will be obvious that a problem exists, but the exact nature of the problem and how to resolve it are not always apparent. In less extreme cases the algorithm will seem to work in that the apparent resolution is enhanced but the results will be unacceptable for other reasons, such as the generation of noise or asymmetric line profiles, or possibly a shifting of spectral features. The programs and discussions of Appendixes 1–4 should give the reader some useful tools with which to gain insight and experience into the benefits *and* drawbacks of deconvolution. We mention several additional sources of potential difficulty, not really hoping to solve anyone's problems before they arise, but hoping mainly to direct the reader's attention to what can be expected from deconvolving spectra of various types under a wide variety of conditions.

Of all the problems that can be encountered in attempting deconvolution, the most serious is the failure of the algorithm to converge. There are a multitude of potential causes for this, but the most common is noise. A separate chapter has been dedicated to this subject, but a summary of one test of the effects of noise on deconvolution is presented here. Figure 24 represents a computer-generated spectrum that has been noise corrupted to produce decreasing signal-to-noise ratios (∞, 100, 50, 20, and 10), prior to deconvolution. Traces (2b), (3b), (4b), (5b), and (6b), respectively, show the results of deconvolution using identical deconvolution parameters. In the lowest signal-to-noise case, trace (6a)–(6b), the algorithm is beginning to diverge, as can be seen in the composite plot of the resulting root-mean-square corrections as a function of iteration number (Fig. 25). A reasonable minimum requirement for deconvolvable data would be a signal-to-noise ratio of 50:1 or greater.

Though noise has clearly been demonstrated to cause the deconvolution algorithm to diverge, any of a number of other problems could manifest themselves in this way. An appropriate choice of α_{max}, the relaxation parameter, is necessary to assure convergence. In general, we have found that for the first few iterations an α_{max} of 0.5 is appropriate. As the root-mean-square corrections decrease, α_{max} may be increased. Unless the data are very good (precise and noise-free), α_{max} should be limited to $\alpha_{max} = 3.0$. (In many of the tests described in this chapter, α_{max} was fixed at a value of $\alpha_{max} = 2.0$ for the purpose of consistency.)

Divergence of the deconvolution algorithm could also be caused by a poorly defined system transfer function or by a zero signal that is not sufficiently well determined. These latter factors, though important, do not in general seriously degrade the results of the deconvolution algorithm; they can put a limit on the resolution enhancement that can be realized. As long as reasonable estimates are made of the transfer function and zero signal level are made, some resolution enhancement can be achieved. The following tests should make clear what is meant by "reasonable estimates."

An accurate characterization of the system transfer function is required if worthwhile deconvolution results are to be obtained. As the following tests reveal, however, this may not be as stringent a requirement as intuition would lead us to believe. In fact it is somewhat surprising how insensitive the results of the deconvolution algorithm are to small changes in the width and shape of the system transfer function.

The results of the first of these tests are illustrated in Figs. 34–36. Using the program DECO described in Appendixes 1 and 2, we have generated an absorption spectrum assuming a Gaussian line width of 3 points and a Gaussian system transfer function with a width of 10 points. Trace (a) of Fig. 34 is the original data and trace (b) is the convolved result that is input to the deconvolution algorithm. Traces (c) through (i) show the results of 50 iterations at $\alpha_{max} = 2.0$ for system transfer functions with Gaussian widths of 7 through 13 points, respectively. The weighting function used is given in Eq. (96) and no interpolation was carried out. Close inspection of the deconvolution results reveals that the proper system transfer function with a full width of 10 points [trace (f)] produces the best results in terms of resolution enhancement. However, the other results are also acceptable in that they *do* converge to physically meaningful results with no generation of additional features or noise; all show some resolution improvements. In fact, the results for system transfer function widths of 9 through 12 points, traces (e) through (h), are hardly distinguishable.

The root-mean-square corrections for the preceding deconvolution tests are plotted in Fig. 35. Since the initial root-mean-square correction is also highly dependent on the data being deconvolved (more lines, greater average correction), it is impossible to test the correctness of an assumed system transfer function by examining the initial root-mean-square correction or the root-mean-square correction as a function of iteration number. For system transfer functions that are too wide, the root-mean-square corrections remain large because the algorithm oscillates about the best possible solution, since there can be no exact solution. In Fig. 36 we have replotted the root-mean-square corrections normalized by dividing by the initial root-mean-square correction. In Fig. 35 and 36, the proper choice for the system transfer function produces the greatest *rate of decrease* of the root-mean-square correction after 50 iterations [trace (d) in both cases]. In addition, the convergence appears to be more uniform in trace (d). This seems to indicate that traces (g) or (e) in Fig. 34, where the

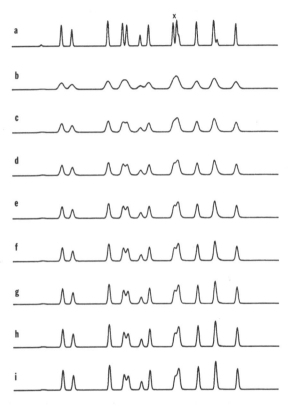

FIG. 34 Test of the effects of varying system transfer function width on deconvolution results using DECO: (a) original spectrum; (b) result of the convolution with a Gaussian 10 points wide; (c, d, e, f, g, h, and i) deconvolution results for an assumed system transfer function 7, 8, 9, 10, 11, 12, and 13 points wide.

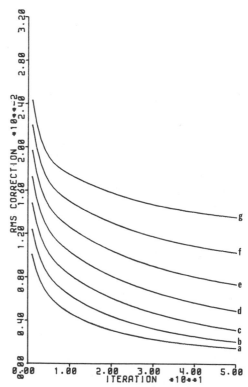

FIG. 35 Composite plot of the root-mean-square corrections for the data and deconvolutions of Fig. 34 plotted against the iteration number; traces (a) through (g) represent widths 7 through 13, respectively.

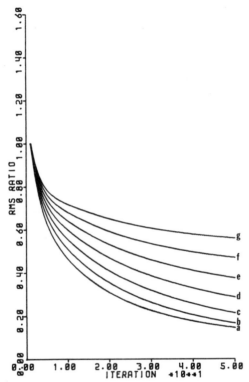

FIG. 36 Normalized root-mean-square corrections for the data of Fig. 34 as a function of iteration number. The root-mean-square corrections were normalized by deciding by the root-mean-square correction from the first iteration; traces (a) through (g) represent widths 7 through 13, respectively.

system transfer function was 10% too large or too small, respectively, would appear less similar to trace (f), where the system transfer function was correct if the number of iterations were increased from 50 to 300.

Also note the feature marked x in the initial data trace (a) of Fig. 34. As the system transfer function used in the deconvolution is widened, the strong–weak doublet is compressed more and more into one strong symmetric line profile. However, this does not seem too severe for a 10% overestimate of the system transfer function width.

A similar test was performed varying the shape of the system transfer function. The computer program DECO was used to generate the same data as in the previous test. The raw data were then convoluted with an asymmetric system transfer function characterized by

$$T(\nu) = T_0 \exp\left[-\frac{(\nu - \nu_0)^2}{2\sigma^2} \right](1 - \beta(\nu - \nu_0)), \qquad (104)$$

where T_0 is a normalization factor and β is the asymmetry factor. The data were then deconvolved with a symmetric Gaussian profile using the same full width (same σ). The results are displayed in Fig. 37 for three cases. Trace (a) is the raw data. Traces (b), (d), and (f) represent the convoluted spectra for $\beta = 0$, 0.1, and 0.2, respectively, using a full width of 10 points. Traces (c), (e), and (g) represent the deconvolved results using a full-width of 10 points and $\beta = 0$ in all cases. Though there is little difference in the convoluted spectra, the deconvolved spectra in Fig. 37 exhibit some effects when the slight asymmetry is ignored in chosing the system transfer function. This is most evident in the doublets resolved by the deconvolution, as one line seems to steal intensity from the other. A very small shifting can be observed as well.

Another potential source of problems in obtaining worthwhile deconvolution results is the estimation of the baseline or zero level. In other spectrometers or for other types of spectral data this may not present any difficulty, since the zero level may be easily determined due to the nature of the experiment or the measurement apparatus. In our case, however, the baseline must be estimated where there are no spectral features and assumed to be continuous and slowly varying (cf. Chapter 10). Thus, we have had an interest in testing the effects of baseline accuracy on deconvolution. One reason for obtaining an accurate zero level is that deconvolution results constrained to non-

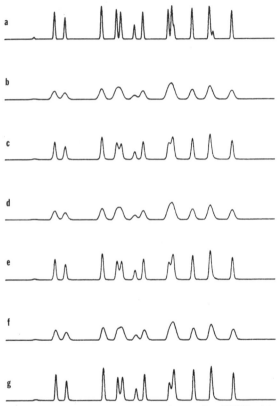

FIG. 37 Test of system transfer function shape using DECO: (a) original spectrum; (b) convolved spectrum $\beta = 0$; (c) deconvolved results; (d) convolved spectrum $\beta = 0.1$; (e) deconvolved results; (f) convolved spectrum $\beta = 0.2$; (g) deconvolved result, where β is the asymmetry parameter as defined in the text. The deconvolutions were performed using a symmetric, $\beta = 0$, Gaussian profile.

negative results, where applicable, are superior to deconvolution results in the unconstrained case.

Figure 38 displays the results of one such test. Trace (a) shows a computer-generated spectrum and trace (b) shows the results of the convolution using a 10-point full-width Gaussian as the system transfer function. Trace (c) shows the deconvolution results for 50 iterations with $\alpha_{max} = 2.0$, using the weighting scheme of Jansson. The results displayed in trace (d) are for the same deconvolution parameters except that an offset of 5% of full scale was added to the convoluted data, trace (b), prior to deconvolution. It is easy to see in trace (d) that spurious features have been generated. Those marked x could easily be mistaken for true spectral features, though they are absent in trace (a). The results displayed in trace (e) were obtained in the same manner except that the offset was -5% of full scale. The features of trace (e) marked y are clearly seen to be reduced in intensity if one examines traces (a) and (e) together. The results displayed in traces (d) and (e) are not without value, however. If one keeps in mind the limits of deconvolution in situations where the baseline is not well known, either of traces (d) or (e) would yield more information than that which would normally be available, that is, in trace (b), the convolved spectrum.

As an extension of the previous test, we now examine the baseline constraint in more detail. This is accomplished using the computer program DECO described in Appendix 1 and listed in Appendix 2. The weight codes (variable KW) used were 0 for no weights (i.e., weight = 1.0) and 1 for the weighting of Willson [10] (weight = $x(1 - x)$, where x is the value of the current deconvolution result at the point to be corrected). An additional test was performed with no weights, except that negative results were reset to zero after each iteration. The results are summarized in Fig. 39. The top two traces (a) and (b) are the same as before. Trace (c) is the result using Willson's weighting scheme ($KW = 1$). Trace (d) is a plot of the results obtained when no weighting was used with negative results set to zero, and trace (e) represents the results for no weights ($KW = 0$). Again 50 iterations with $\alpha_{max} = 2.0$ were used for all tests. In Fig. 40, the root-mean-square correction for each deconvolution test is plotted. Traces (b) and (c), which correspond to traces (d) and (e) of Fig. 39, exhibit a much greater decrease in the root-mean-square correction, because of the greater degree of freedom for the corresponding weighting schemes.

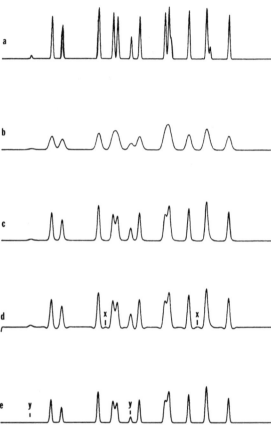

FIG. 38 Test of baseline or zero-level accuracy on deconvolution. Trace (a) is a computer-generated spectrum and trace (b) is the result of the convolution of the spectrum with a 10-point Gaussian. Trace (c) is the deconvolution results for 50 iterations with $\alpha_{max} = 2$. Trace (d) and trace (e) are the deconvolution results after an offset of $+5\%$ and -5% of full scale, respectively, was added to trace (b) before deconvolution. Features marked x and y show intensity variations due to inaccurate baseline removal.

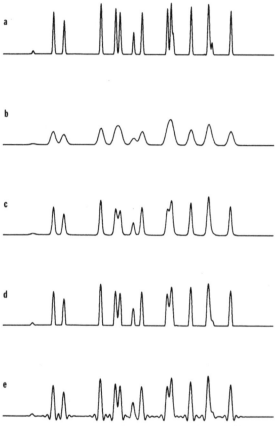

FIG. 39 Effects of nonnegativity constraints. Trace (a) is the original spectrum, (b) the convolved spectrum for a system transfer function 10 points wide, (c) the deconvolved spectrum using Jansson's constraining method, (d) the deconvolved spectrum using no constraints but setting negative results to zero after each iteration, and (e) the unconstrained results.

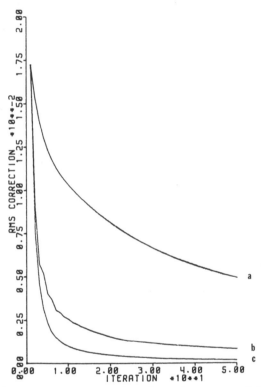

FIG. 40 Root-mean-square corrections for various nonnegativity constraints. Traces (a), (b), and (c) correspond to the data of Fig. 39 for the results displayed in traces (c), (d), and (e), respectively.

Clearly, trace (e) of Fig. 39 is unacceptable for data that have a nonnegativity constraint imposed by the nature of the experiment. Also, many features are generated that appear as side lobes to the stronger lines. These side lobes do not correspond to features in the original spectrum, trace (a). Whether trace (c) or trace (d) of Fig. 39 represents the better result depends on the demands of the experiment. Both show marked resolution enhancement, good stability, and no generation of artifacts that could be confused with genuine lines. It is possible that other weighting schemes would be optimum for other types of spectra.

A problem of a different nature that is likely to arise when using deconvolution is the consumption of large amounts of computer time due to the iterative nature of the deconvolution algorithm. There are three things to consider if the amount of computer time becomes prohibitive. One consideration is the data point density or the number of data points recorded in a fixed time or frequency interval. The time required to deconvolve a one-wavenumber segment is more or less proportional to the square of the data point density. Thus, if the number of data points corresponding to the full width of a single, isolated line can be reduced from 30 to 15 by dropping every other point, a savings of 75% in computer time will be realized, provided 15 data points are sufficient to characterize the line profile. Note that a 15-point/FWHM density may correspond to only 5 data points or less for the full width of the same line after deconvolution. Another way to reduce computer time is to use interpolation during the initial iterations. If in each iteration, the correction is computed at every Kth data point, and interpolation is used to find the corrections at the intervening data points, the computer time required will be reduced by a factor of nearly K. If, for example, the correction is calculated at every third point, i.e., at points 3, 6, 9, 12, . . . , the interpolation is carried out in such a way that $C_4 = \frac{2}{3} C_3 + \frac{1}{3} C_6$, etc. Formally, if corrections are calculated at every k step, then the corrections are given by

$$ C_{ki+j} = \left[\left(\frac{k-j}{k} \right) C_{ki} + \frac{j}{k} C_{k(i+1)} \right], \qquad j = 0, 1, 2, \ldots, k - 1. \quad (105) $$

Third, a large percentage of the computer time is consumed by the calculation of the convolution of the current result and the system transfer function at each iteration. If this section of the computer program (subroutine CONV of Appendix 2) is rewritten in a lower-level language rather than FORTRAN, a considerable time savings can be realized (60% on an IBM 360/65, for example).

Since much of the success of deconvolution depends on determining the system transfer function, we discuss this in more detail. Generally, the preferred method of determining the system transfer function is to look at a single, isolated absorption line, under pressure and temperature conditions designed to eliminate any pressure-induced effects in the line profile. Then the measured line shape S can be written as

$$S(t) = T(t,\nu) \oplus D(\nu), \qquad (106)$$

where $T(t,\nu)$ is the system transfer function and $D(\nu)$ is the known line shape, in our case generally a Doppler-broadened profile. Since S and D are known, we can use the deconvolution process and programs to find T by reversing the roles of D and T. In solving for T it is desirable to have a very high signal-to-noise ratio, but it is also necessary to scan the line with the same parameters that are used to scan the data to be deconvolved. Most cases would require system parameters such that it would require a prohibitive amount of time to record a spectrum. We avoid this problem by selecting scanning parameters that will give a suitable signal-to-noise ratio in the data to be deconvolved. Using these same parameters we scan an isolated line many times and average the multiple scans to reduce the noise in the instrument function profile to an acceptable level. A good treatment of this procedure is given by Jansson [24].

On many occasions it has not been practical to redetermine T for each data record, and we have often resorted to using a Gaussian approximation. That is, since our instrument transfer function is quite closely approximated by a Gaussian function, we simply assume that it is Gaussian and proceed as follows. Our data are most often composed of Doppler-limited absorption lines. The resulting observed transitions are therefore nearly Gaussian with a width equal to the square root of the sum of the squares of the widths of the instrument function and the Doppler width. It is thus a simple matter to examine the data and estimate the appropriate width to use for the assumed Gaussian instrument function.

This seems to yield a reasonable approximation to the system transfer function, as we have routinely achieved resolution enhancements of 2.5–3. In experimenting with various system transfer functions approximations, we have found that the width is the most critical feature and that details of the system transfer function profile are slightly less critical. This may be the case because our spectrometer produces nearly symmetric, Gaussian profiles. Also, we should point out, that in our experimental set-up, the system transfer function remains essentially the same throughout a data scan. If this is not the case some adjustments will have to be made. One such possibility we have considered is to transform the independent variable (time, frequency, etc.) so that the system transfer function will not depend on the independent variable, deconvolving, and then reversing the transformation. In any case, for experimental set-ups much different than that which we have described, some ingenuity may be needed to obtain a satisfactory approximation to the system transfer function.

CHAPTER 13

ASSESSMENT AND EVALUATION

The project that ultimately gave birth to this monograph started out as an attempt to test the Jansson modification [8] of the Van Cittert [18] deconvolution algorithm. Although Jansson's contribution is often referred to as a modification of Van Cittert's work, it would be more appropriate simply to speak of Jansson's algorithm, since it was his inclusion of overrelaxation techniques that produced a useful and stable deconvolution algorithm (cf. Chapter 5).

We have found that the positional stability of spectral features under deconvolution is excellent. The tests discussed in Chapter 8 indicate this; moreover, actual use of deconvolution in our laboratory has verified these early tests. Deconvolution carried out properly and carefully does not shift the frequency of spectral features in any measureable way. Whatever statistical scatter we currently observe in measured transition frequencies appears to be the result of instrumental limitations and not the deconvolution algorithm. M. A. Dakhil, working in our laboratory, has observed hundreds of thousands of calibrated spectral frequencies in the $^{12}CD_3F$ spectrum at 5-μm wavelength. He has calibrated and measured these frequencies in both nondeconvolved and deconvolved data sets and no significant shifts have been observed.

It is true that certain implementations of the Jansson algorithm may result in positional instabilities, and these are discussed by Willson [10].

Preliminary tests as discussed in Chapter 11 indicate that recovery of intensity information under deconvolution is quite good. Intensity tests, however, are much more difficult to carry out on real data than are tests of frequency stability. The results in Chapter 11 indicate that deconvolution may well prove to be a useful tool for intensity studies of relatively dense, Doppler-limited spectra. Other preliminary studies of line profile recovery have yielded promising results where the instrument function width is as large as 50% of the true line width. Several tests using a Gaussian instrument function that has a width equal to one-half the width of a Lorentzian test line indicate that profile recovery is possible, especially in the wings of the Lorentzian line.

The experienced spectroscopist may have found fault with the discussion in Chapter 1 on the benefits of resolution degradation followed by resolution recovery using deconvolution. The discussion did not include any consideration of noise and is, to that extent, flawed. However, the sense of the results is in fact born out in practice. In the area of high-resolution precision measurement, nature provides seemingly unalterable difficulties especially as regards the trade-off of resolution versus signal-to-noise ratio. At first glance, the idea of opening the slits of a spectrometer so as to allow a broader-frequency passband to be used and thus a higher data acquisition rate does not seem realistic. One would conclude that using deconvolution to recover the lost resolution would simultaneously "recover" the "lost" noise. In fact, this is not the case. Because Jansson's algorithm is nonlinear (cf. Chapter 5), typical "linear" intuition fails. Since noise is particularly bothersome at 0% absorption (it may be confused with weakly absorbing features) and since corrections to iteration j are multiplied by the damping factor $\alpha_n(L_{n-1}(t))$ (cf. Chapter 6), which is zero at 0% absorption and very small at 10% absorption compared to $\alpha_{n \, max}$ at 50% absorption, noise is not enhanced at anywhere near the rate (per iteration) that resolution is enhanced.

In Chapter 1, we show that for a constant signal-to-noise ratio the scanning rate varies as the fifth power of the ratio of the resolution. If the signal-to-noise ratio in the degraded resolution data run must be n times greater than the nondeconvolved nondegraded data run to achieve an acceptable signal-to-noise ratio in the deconvolved spectrum, Eq. (32) becomes

$$\frac{R_1}{R_0} < \left(\frac{1}{n^2}\right)\left(\frac{W_1}{W_0}\right)^5. \tag{107}$$

We find that deconvolution resulting in a resolution enhancement of a factor of 3 does not enhance noise by more than a factor of 1.5–2. Thus, to maintain equal signal-to-noise ratios, the scanning rate enhancement of 243 in Table I must in practice be divided by 2.25 to 4, yielding a scan rate enhancement range of 108–60.75, respectively. For deconvolution that results in resolution enhancements of 2 or less, the noise enhancement is essentially negligible. All these results are predicated on a signal-to-noise ratio in the raw data of 50:1 to 150:1.

Interestingly enough, it seems to be the actual signal-to-noise ratio that is significant and not the information-to-noise ratio. That is, the same considerations seem to apply for a spectrum where maximum apparent absorption is 20% or 88%.

Overall, we have found deconvolution to be a useful, stable technique when carefully used and results checked for reasonableness. On the other hand, one must always remember that the process seeks to enhance lost (attenuated) high spectral frequencies (in the sense of the Fourier transform). Thus, data that are subjected to processing that injects alien high spectral frequencies not consistent with the conditions under which the data were obtained may cause invalid and/or unpredictable results.

USING THE PROGRAM DECO

Listed in Appendix 2 is the computer program that was used for many of the deconvolution tests we have performed. It is designed to simulate data of various descriptions, convolve the data with various types of system transfer functions, and then perform the deconvolution algorithms described earlier. The program allows one to examine the results at any stage of deconvolution in one of several ways. We have at our disposal a small incremental plotter for which subroutines similar to the familiar CALCOMP routines have been written. This allows us to use the program DECO interactively. That is, we can examine intermediate results before proceeding further.

The computer system used for these tests consists of a Digital Equipment Corporation PDP 11/20 with 28 kilowords of core memory, a teletype, a Houston Instruments/COMPLOT digital plotter with 200 points per inch resolution, and a moving head disk with 500 kilobytes of storage space. The operating system used was RT11 version 2 and FORTRAN version 01-11, both supplied by Digital Equipment Corporation. All computer programs used in these tests are written in FORTRAN.

The first section of the program is the data generation section, beginning at line 7. The variables NL and AM determine how the data are to be generated. The options are as follows: for NL positive, NL random line positions with strengths from 0 to AM, for NL equal zero the entire spectrum is read in from unit 1, and for NL negative, $-NL$ positions and strengths are read from unit 5, the teletype in our case. At this point the spectrum consists of infinitely narrow lines sometimes called a stick spectrum. Next, the stick spectrum is convolved with a line shape, which is either read in for LTYP = 0, generated as a Gaussian for LTYP = 1, or generated as a Lorentzian for LTYP = 2. The variable FWHM controls the width of the generated line for LTYP = 1 or 2. The array F is used to store the line shape temporarily. After the line shape is convolved in, the data can be converted to absorption if the next input variable (K) is zero. Otherwise, the data are directly scaled from 0 to 1.

The response function is generated in the same manner as the line shape and is stored in array F. The data are plotted, convolved with the response function, and plotted again.

The next input controls the noise generation section of the program. For NTYP = 0, no noise is generated. The variable AM controls the magnitude of the noise for NTYP not zero. For NTYP = 1, an even distribution from $-AM$ to $+AM$ is generated. For NTYP = 2, a Gaussian distribution with a standard deviation of AM is generated and for NTYP = 2, the noise generated is proportional to the square root of the data array at each point.

This brings us to the deconvolution section starting at line 77, where the deconvolution control parameters are read. For LP positive, LP deconvolution iterations are performed, with ALF as α_{max}, KK as the interpolation interval, and KW as the weight code. These will be explained later. For $LP = -9$, negative points are set to 0. For $LP = -10$, a new data set is generated. For LP from -8 to -1, plotting is performed. There are four arrays that can be plotted: the current results, the original spectrum, the convolved spectrum, and the error spectrum, which is the algebraic difference between the original spectrum and the current result. Also, the power spectral density, phase spectrum, or both can be plotted for each array for KK positive, negative, or zero, respectively. This is done in subroutine PSD.

The weight code KW is used to control the form of the weighting function. This is used in routine WT. For $KW = 0$, no weighting is used. For $KW = 1$, the weighting scheme of Willson is used [10]. For $KW = 2$, a weighting scheme for data with a lower bound of 0 and no upper bound is used.

The interpolation interval, KK, is set equal to 1 for normal deconvolution. However, in order to save computer time, the corrections can be interpolated between every KKth data point. For example, if $KK = 3$, the corrections are computed directly for every third data point and two thirds of the correction is applied at the neighboring points and one third at data points two elements away.

Input flow:

Card #	Variables	Format	
1	NL, AM	I5, F5.0	
2 → REPEAT-NL TIMES IF NL < 0	K, AM	I5, F5.0	for NL negative
3	LTYP, FWHM	I5, F5.0	for line profile
4	LTYP, FWHM	I5, F5.0	for response function
5	NTYP, AM	I5, F5.0	for noise generation
6	NS	I5	smoothing interval
7 → REPEATED UNTIL LP = 0	LP, ALF, KK, KW	I5, F5.0, I5, I5	deconvolution/plotting parameters

APPENDIX 2
DECO

```
       C --- DECO.FOR
       C --- DECONVOLUTION TEST 12-APR-77
       C --- A = DATA ARRAY
       C --- F = RESPONSE FUNCTION
       C --- CC,D,Q SECONDARY DATA ARRAYS
       C --- D = CURRENT DECONVOLUTED RESULT IN DECON SECTION
       C --- IBUF = PLOTTER STORAGE
0001           DIMENSION A(512),D(512),F(200),IBUF(500),CC(512)
0002           DIMENSION FMT(20),Q(512)
0003           COMPLEX SS(512),COSSIN(9)
       C --- INITIALIZE PLOTTER; CALCOMP STANDARD ROUTINES USED.
0004           CALL PLOTS(IBUF,500,120.)
0005           MD=9
0006           ND=512
       C
       C --- DATA GENERATION SECTION
       C
       C --- ZERO DATA ARRAYS
0007   5000    CONTINUE
0008           CALL PLOT(3.,-2.,-3)
0009           CALL PLOT(0.,0.5,-3)
0010           DO 30 I=1,ND
0011           D(I)=0.
0012           CC(I)=0.
0013   30      A(I)=0.
       C --- NL = # OF LINES TO GENERATE
       C     NL > 0  GENERATE NL (RANDOM) LINE POSITIONS AND INTENSITIES
       C     NL = 0  READ DATA FILE (UNIT 1)
       C     NL < 0  READ NL POSITIONS AND INTENSITIES
       C --- AM = MAXIMUM LINE TO GENERATE
0014           READ(5,101) NL,AM
0015           IF(NL.LT.0) GO TO 50
0017           IF(NL.EQ.0) GO TO 53
0019           IR1=0
0020           IR2=0
0021           DO 1 I=1,NL
0022           K=RAN(IR1,IR2)*(ND-100.)+50.
0023   1       A(K)=RAN(IR1,IR2)*AM
0024           GO TO 52
       C --- READ -NL LINE POSITIONS AND STRENGTHS
0025   50      NL=-NL
0026           DO 51 I=1,NL
0027           READ(5,101) K,AM
0028   51      A(K)=AM
0029           GO TO 52
0030   53      READ(1,300) FMT
0031   300     FORMAT(20A4)
0032           READ(1,FMT) (A(I),I 1,ND)
       C
       C
0033   52      READ(5,101) LTYP,FWHM
       C --- LTYP=0, READ LINE SHAPE
       C --- LTYP=+1, GAUSSIAN LINE SHAPE
       C --- LTYP=+2, LORENTZIAN LINE SHAPE
       C --- USE F TO STORE LINE SHAPE
0034           CALL GENER(LTYP,F,NF,FWHM,IX)
       C --- CONVOLUTE WITH LINE SHAPE INTO D
```

```
0035            CALL CONV(A,ND,F,NF,IX,D,1)
         C --- STORE BACK INTO A
0036            DO 142 I=1,ND
0037     142    A(I)=D(I)
         C --- CONVERT TO ABSORPTION IF DESIRED (K NOT ZERO)
         C --- SET MAX = 1.0 OTHERWISE
0038            READ(5,101) K
0039            IF(K.EQ.0) GO TO 120
0041            DO 121 I=1,ND
0042     121    A(I)=1.-EXP(-A(I))
0043            GO TO 122
         C --- SET MAX = 1.0
0044     120    AM=0.
0045            DO 123 I=1,ND
0046     123    IF(AM.LT.A(I)) AM=A(I)
0048            DO 124 I=1,ND
0049     124    A(I)=A(I)/AM
         C --- GENERATE RESPONSE FUNCTION
0050     122    READ(5,101) LTYP,FWHM
0051            CALL GENER(LTYP,F,NF,FWHM,IX1)
         C --- STORE DATA IN CC
0052            DO 43 I=1,ND
0053     43     CC(I)=A(I)
0054            CALL LOOK(CC,ND,0.,3)
         C --- CONVOLUTE WITH RESPONSE FUNCTION BACK INTO A
0055            CALL CONV(CC,ND,F,NF,IX1,A,1)
         C --- PLOT RESULT
0056            CALL LOOK(A,ND,0.,7)
         C
         C --- NOISE ADDITION SECTION
         C --- GENERATE NOISE TO ADD TO CONVOLUTED SPECTRUM
         C
         C      NTYP = 0   NO NOISE, SKIP TO NEXT SECTION
         C      NTYP = 1   EVEN DISTRIBUTION OF NOISE
         C      NTYP = 2   GAUSSIAN DISTRIBUTION OF NOISE
         C      NTYP = 3   NOISE PROPORTIONAL TO SQRT OF SIGNAL
         C
         C      AM CONTROLS SPREAD OF NOISE OR SIGNAL TO NOISE RATIO
         C
0057            READ(5,101) NTYP,AM
0058            IF(NTYP.EQ.0) GO TO 60
0060            IF(NTYP.NE.1) GO TO 61
0062            DO 62 I=1,ND
0063     62     A(I)=A(I)+2.*(.5-RAN(IR1,IR2))*AM
0064            GO TO 60
0065     61     IF(NTYP.NE.2) GO TO 63
0067            DO 64 I=1,ND
0068     64     A(I)=A(I)+(.5-GAUS(IR1,IR2))*AM
0069            GO TO 60
0070     63     IF(NTYP.NE.3) GO TO 60
0072            DO 65 I=1,ND
0073     65     A(I)=A(I)+(.5-RAN(IR1,IR2))*2.*AM*SQRT(A(I))
0074     60     CONTINUE
         C
         C --- SMOOTH DATA ARRAY FOR NS > 0
         C
0075            READ(5,101) NS
```

```
0076          IF(NS.EQ.0) GO TO 90
       C --- COPY A INTO D
0078          DO 91 I=1,ND
0079   91     D(I)=A(I)
       C --- SMOOTH D INTO A
0080          CALL SMOOTH(D,ND,NS,A)
0081   90     CONTINUE
       C
       C --- DECONVOLUTION SECTION
       C
       C --- INITIALIZE D=A
0082          DO 2 I=1,ND
0083   2      D(I)=A(I)
       C --- READ CONTROL CARD
0084   10     READ(5,101) LP,ALF,KK,KW
0085   101    FORMAT(I5,F5.0,I5,I5)
0086          IF(LP.LE.0)GOTO301
0088          AM=LP
0089          CALL PLOT(1.,0.,-3)
0090          CALL SYMBOL(0.,0.,0.14,'ITERATE:',90.,8)
0091          CALL NUMBER(999.,999.,0.14,AM,90.,0)
0092          CALL SYMBOL(999.,999.,0.14,'ALPHA=   ',90.,8)
0093          CALL NUMBER(999.,999.,0.14,ALF,90.,2)
0094          CALL PLOT(0.,0.,3)
       C --- LP:
       C        + LP DECONVOLUTION ITERATIONS
       C        0 EXIT
       C       -1 PLOT CURRENT RESULT (D)
       C       -2 PLOT POWER SPECTRUM OF CURRENT RESULT
       C       -3 PLOT ORIGINAL SPECTRUM (CC)
       C       -4 PLOT POWER SPECTRUM OF ORIGINAL
       C       -5 PLOT DIFFERENCE OF CURRENT RESULT AND DESIRED RESULT
       C       -6 PLOT POWER SPECTRUM OF DIFFERENCE
       C       -7 PLOT CONVOLUTED SPECTRUM (MEASURED SPECTRUM)
       C       -8 PLOT POWER SPECTRUM OF CONVOLUTED SPECTRUM
       C       -9 SET POINTS <0 TO 0
       C      -10 GET NEXT CASE DATA SET
       C
       C --- ALF:
       C        DECONVOLUTION WEIGHTING FACTOR FOR LP>0
       C        PLOT SIZE FOR LP<0
       C
0095   301    CONTINUE
0096          IF(LP.GT.0) GO TO 78
0098          IF(LP.EQ.0) GO TO 77
0100          LP=-LP
0101          GO TO (76,75,74,73,72,71,70,69,68,5000),LP
0102          GO TO 77
0103   73     SF=0.
0104          DO 20 I=KK,NF,KK
0105   20     SF=SF+F(I)
0106          ND2=ND-NF+IX1
0107          K2=KK-1
0108          DO 21 L=1,LP
       C --- FORM CONVOLUTION OF CURRENT RESULT IN Q
0109          CALL CONV(D,ND,F,NF,IX1,Q,KK)
       C --- CORRECT CURRENT APPROXAMATION
```

```
0110            SD=0.
0111            DO 26 IX=IX1,ND2,KK
0112            DEV=A(IX)-Q(IX)/SF
0113            SD=SD+DEV**2
     C --- WEIGHT CORRECTION
0114            DEV=DEV*WT(KW,D(IX))*ALF
     C --- APPLY CORRECTION, INTERPOLATING FOR KK>1
0115            IF(KK.EQ.1) GO TO 26
0117            K2=KK-1
0118            DO 25 K=1,K2
0119            D(IX+K)=D(IX+K)+DEV*K/KK
0120    25      D(IX-K)=D(IX-K)+DEV*K/KK
0121    26      D(IX)=D(IX)+DEV
0122            SD SQRT(SD/ND*KK)
0123    21      WRITE(6,200) L,SD
0124    200     FORMAT(I4,1X,E12.4)
0125            GO TO 10
0126    76      CALL LOOK(D,ND,ALF,LP)
0127            GO TO 10
0128    75      CALL PSD(D,ND,MD,COSSIN,SS,ALF,KK,LP)
0129            GO TO 10
0130    74      CALL LOOK(CC,ND,ALF,LP)
0131            GO TO 10
0132    73      CALL PSD(CC,ND,MD,COSSIN,SS,ALF,KK,LP)
0133            GO TO 10
0134    72      DO 172 I=1,ND
0135    172     Q(I)=CC(I)-D(I)
0136            CALL LOOK(Q,ND,ALF,LP)
0137            GO TO 10
0138    71      DO 171 I=1,ND
0139    171     Q(I)=CC(I)-D(I)
0140            CALL PSD(Q,ND,MD,COSSIN,SS,ALF,KK,LP)
0141            GO TO 10
0142    70      CALL LOOK(A,ND,ALF,LP)
0143            GO TO 10
0144    69      CALL PSD(A,ND,MD,COSSIN,SS,ALF,KK,LP)
0145            GO TO 10
0146    68      DO 168 I=1,ND
0147    168     IF(D(I).LT.0.) D(I)=0.
0149            GO TO 10
0150    77      CALL PLOT(0.,0.,999)
0151            CALL EXIT
0152            END
```

RT-11 FORTRAN IV   STORAGE MAP

NAME	OFFSET	ATTRIBUTES	
A	000006	REAL*4	ARRAY (512)
D	004006	REAL*4	ARRAY (512)
F	010006	REAL*4	ARRAY (200)
IBUF	011446	INTEGER*2	ARRAY (500)
CC	013416	REAL*4	ARRAY (512)
FMT	017416	REAL*4	ARRAY (20)
Q	017536	REAL*4	ARRAY (512)
SS	023536	COMPLEX*3	ARRAY (512)
COSSIN	033536	COMPLEX*8	ARRAY (9)
PLOTS	000000	REAL*4	PROCEDURE
MD	034004	INTEGER*2	VARIABLE
ND	034006	INTEGER*2	VARIABLE
PLOT	000000	REAL*4	PROCEDURE
I	034010	INTEGER*2	VARIABLE
NL	034012	INTEGER*2	VARIABLE
AM	034014	REAL*4	VARIABLE
IR1	034020	INTEGER*2	VARIABLE
IR2	034022	INTEGER*2	VARIABLE
K	034024	INTEGER*2	VARIABLE
RAN	000000	REAL*4	PROCEDURE
LTYP	034026	INTEGER*2	VARIABLE
FWHM	034030	REAL*4	VARIABLE
GENER	000000	REAL*4	PROCEDURE
NF	034034	INTEGER*2	VARIABLE
IX	034036	INTEGER*2	VARIABLE
CONV	000000	REAL*4	PROCEDURE
EXP	000000	REAL*4	PROCEDURE
IX1	034040	INTEGER*2	VARIABLE
LOOK	000000	INTEGER*2	PROCEDURE
NTYP	034042	INTEGER*2	VARIABLE
GAUS	000000	REAL*4	PROCEDURE
SQRT	000000	REAL*4	PROCEDURE
NS	034044	INTEGER*2	VARIABLE
SMOOTH	000000	REAL*4	PROCEDURE
LP	034046	INTEGER*2	VARIABLE
ALF	034050	REAL*4	VARIABLE
KK	034054	INTEGER*2	VARIABLE
KW	034056	INTEGER*2	VARIABLE
SYMBOL	000000	REAL*4	PROCEDURE
NUMBER	000000	INTEGER*2	PROCEDURE
SF	034060	REAL*4	VARIABLE
ND2	034064	INTEGER*2	VARIABLE
K2	034066	INTEGER*2	VARIABLE
L	034070	INTEGER*2	VARIABLE
SD	034072	REAL*4	VARIABLE
DEV	034076	REAL*4	VARIABLE
WT	000000	REAL*4	PROCEDURE
PSD	000000	REAL*4	PROCEDURE
EXIT	000000	REAL*4	PROCEDURE

```
        C --- CONVOLUTION SUBROUTINE
        C --- CONVOLUTE X WITH F
        C --- STORE RESULT IN Y
        C --- IX IS 'CENTER' OF F
        C --- NF = # OF POINTS IN F
        C --- NX = # OF POINTS IN X,Y
0001            SUBROUTINE CONV(X,NX,F,NF,IX,Y,KK)
0002            DIMENSION X(NX),Y(NX),F(NF)
0003            N=NX-NF+IX
0004            DO 1 I=IX,N,KK
0005            S=0.
0006            DO 2 J=1,NF
0007    2       S=S+F(J)*X(I+J-IX)
0008    1       Y(I)=S
0009            RETURN
0010            END
```

RT-11 FORTRAN IV   STORAGE MAP

NAME	OFFSET	ATTRIBUTES		
X	000014	REAL*4	PARAMETER	ARRAY (NX)
NX	000016	INTEGER*2	PARAMETER	VARIABLE
Y	000026	REAL*4	PARAMETER	ARRAY (NX)
F	000020	REAL*4	PARAMETER	ARRAY (NF)
NF	000022	INTEGER*2	PARAMETER	VARIABLE
IX	000024	INTEGER*2	PARAMETER	VARIABLE
KK	000030	INTEGER*2	PARAMETER	VARIABLE
N	000032	INTEGER*2	VARIABLE	
I	000034	INTEGER*2	VARIABLE	
S	000036	REAL*4	VARIABLE	
J	000042	INTEGER*2	VARIABLE	

```
          C --- PLOT DATA TO 10 INCHES
          C --- SCL = 0  DATA IS SCALED TO 1 INCH
          C --- SCL > 0  DATA IS TO BE SCALED TO SCL INCHES
0001          SUBROUTINE LOOK(B,NB,SCL,LP)
0002          REAL*8 TITLE(8)
0003          DIMENSION B(NB)
0004          DATA TITLE/'DCN SP ','PS DCNV','ORIG SP ',
             1'PS: ORIG','ERROR SP','PS: ERROR','CONV SP ',
             2'PS: CONV'/
0005          B1=0.
0006          S=1.
0007          IF(SCL.LE.0.) GO TO 10
0009          B1=B(1)
0010          B2=B1
0011          DO 11 I=1,NB
0012          IF(B(I).LT.B1) B1=B(I)
0014          IF(B(I).GT.B2) B2=B(I)
0016    11    CONTINUE
0017          S=S/(B2-B1)*SCL
0018    10    CALL PLOT(.25,0.,-3)
0019          CALL SYMBOL(0.,0.,0.14,TITLE(LP),90.,8)
0020          IF(SCL.LE.0.) CALL PLOT(S,0.,-3)
0022          IF(SCL.GT.0) CALL PLOT(SCL,0.,-3)
0024          DO 1 I=1,NB
0025     1    CALL PLOT(-(B(I)-B1)*S,I*10./NB,2)
0026          CALL PLOT(0.,0.,3)
0027          RETURN
0028          END
```

RT-11 FORTRAN IV   STORAGE MAP

NAME	OFFSET	ATTRIBUTES	
TITLE	000024	REAL*8	ARRAY (8)
B	000014	REAL*4	PARAMETER ARRAY (NB)
NB	000016	INTEGER*2	PARAMETER VARIABLE
SCL	000020	REAL*4	PARAMETER VARIABLE
LP	000022	INTEGER*2	PARAMETER VARIABLE
B1	000152	REAL*4	VARIABLE
S	000156	REAL*4	VARIABLE
B2	000162	REAL*4	VARIABLE
I	000166	INTEGER*2	VARIABLE
PLOT	000000	REAL*4	PROCEDURE
SYMBOL	000000	REAL*4	PROCEDURE

RT-11 FORTRAN IV   V01-11 SOURCE LISTING   PAGE 001

```
        C --- CALCULATE POWER SPECTRAL DENSITY OR PHASE SPECTRUM OF
ARRAY A
        C --- USES PHASE AND POWER FUNCTIONS
        C --- FFT,CSTABL TO CALCULATE FOURIER TRANSFORM
        C --- KD>=0 PLOT POWER SPECTRUM
        C --- KD<=0 PLOT PHASE SPECTRUM
        C --- AMP IS PLOT SCALE FACTOR
0001          SUBROUTINE PSD(A,ND,MD,COSSIN,SS,AMP,KD,LP)
0002          COMPLEX SS(ND),COSSIN(MD)
0003          DIMENSION A(ND),Q(256)
0004          CALL CSTABL(COSSIN,MD)
0005          DO 1 I=1,ND
0006   1      SS(I)=A(I)
0007          CALL FFT(SS,MD,1,COSSIN)
0008          NQ=ND/2
0009          IF(KD.LT.0) GO TO 2
0011          DO 3 I=1,NQ
0012   3      Q(I)=CABS(SS(I))
0013          CALL LOOK(Q,NQ,AMP,LP)
0014   2      IF(KD.GT.0) GO TO 4
0016          DO 5 I=1,NQ
0017   5      Q(I)=ATAN2(REAL(SS(I)),AIMAG(SS(I)))
0018          CALL LOOK(Q,NQ,AMP,LP)
0019   4      RETURN
0020          END
```

RT-11 FORTRAN IV   STORAGE MAP

NAME	OFFSET	ATTRIBUTES		
SS	000024	COMPLEX*8	PARAMETER ARRAY (ND)	
ND	000016	INTEGER*2	PARAMETER VARIABLE	
COSSIN	000022	COMPLEX*8	PARAMETER ARRAY (MD)	
MD	000020	INTEGER*2	PARAMETER VARIABLE	
A	000014	REAL*4	PARAMETER ARRAY (ND)	
Q	000034	REAL*4	ARRAY (256)	
AMP	000026	REAL*4	PARAMETER VARIABLE	
KD	000030	INTEGER*2	PARAMETER VARIABLE	
LP	000032	INTEGER*2	PARAMETER VARIABLE	
CSTABL	000000	REAL*4	PROCEDURE	
I	002036	INTEGER*2	VARIABLE	
FFT	000000	REAL*4	PROCEDURE	
NQ	002040	INTEGER*2	VARIABLE	
CABS	000000	COMPLEX*8	PROCEDURE	
LOOK	000000	INTEGER*2	PROCEDURE	
ATAN2	000000	REAL*4	PROCEDURE	
REAL	000000	REAL*4	PROCEDURE	
AIMAG	000000	REAL*4	PROCEDURE	

```
        C --- FFT.FOR
0001            SUBROUTINE FFT(A,M,MODE,COSSIN)
        C --- A = COMPLEX*8 INPUT AND RETURNED AS COMPLEX*8 FOURIER
TRANSFORM
        C --- COSSIN IS COS,SIN TABLE  SEE CSTBL ROUTINE
        C --- NUMBER OF POINTS = N = 2**M
        C --- MODE = 1 ==> TRANSFORM; -1 ==> INVERSE TRANSFORM
0002            COMPLEX*8 A(1),U,W,T,COSSIN(1)
0003            N=2**M
0004            ND2=N/2
0005            NM1=N-1
0006            J=1
0007            DO 30 I=1,NM1
0008            IF(I.GE.J) GO TO 10
0010            T=A(J)
0011            A(J)=A(I)
0012            A(I)=T
0013    10      K=ND2
0014    20      IF(K.GE.J) GO TO 30
0016            J=J-K
0017            K=K/2
0018            GO TO 20
0019    30      J=J+K
0020            LP2=1
0021            DO 50 L=1,M
0022            LL=LP2
0023            LP2=LP2*2
0024            U=(1.,0.)
0025            W=COSSIN(L)
0026            IF(MODE.EQ.-1)  W=CONJG(W)
0028            DO 50 J=1,LL
0029            DO 40 I=J,N,LP2
0030            IL=I+LL
0031            T=A(IL)*U
0032            A(IL)=A(I)-T
0033    40      A(I)=A(I)+T
0034    50      U=U*W
0035            IF(MODE.EQ.1)  RETURN
        C --- NORMALIZE INVERSE IF REQUIRED
0037            T=1./N
0038            DO  15  I=1,N
0039    15      A(I)=A(I)*T
0040            RETURN
0041            END
```

RT-11 FORTRAN IV   STORAGE MAP

```
NAME      OFFSET  ATTRIBUTES

A         000014  COMPLEX*8 PARAMETER ARRAY (1)
COSSIN    000022  COMPLEX*8 PARAMETER ARRAY (1)
M         000016  INTEGER*2 PARAMETER VARIABLE
MODE      000020  INTEGER*2 PARAMETER VARIABLE
U         000024  COMPLEX*8 VARIABLE
W         000034  COMPLEX*8 VARIABLE
T         000044  COMPLEX*8 VARIABLE
N         000054  INTEGER*2 VARIABLE
ND2       000056  INTEGER*2 VARIABLE
NM1       000060  INTEGER*2 VARIABLE
J         000062  INTEGER*2 VARIABLE
I         000064  INTEGER*2 VARIABLE
K         000066  INTEGER*2 VARIABLE
LP2       000070  INTEGER*2 VARIABLE
L         000072  INTEGER*2 VARIABLE
LL        000074  INTEGER*2 VARIABLE
CONJG     000000  COMPLEX*8 PROCEDURE
IL        000076  INTEGER*2 VARIABLE
```

RT-11 FORTRAN IV   V01-11 SOURCE LISTING   PAGE 001

```
        C --- CSTABL.FOR
0001          SUBROUTINE CSTABL(COSSIN,M)
        C --- GENERATE COS,SIN TABLE FOR ROUTINE FFT
0002          COMPLEX*8 COSSIN(1),CMPLX
0003          REAL*4 X
0004          X=3.1415926
0005          DO 10 I=1,M
0006          COSSIN(I)=CMPLX(COS(X),SIN(X))
0007       10 X=X/2.
0008          RETURN
0009          END
```

RT-11 FORTRAN IV   STORAGE MAP

```
NAME      OFFSET  ATTRIBUTES

COSSIN    000014  COMPLEX*8 PARAMETER ARRAY (1)
M         000016  INTEGER*2 PARAMETER VARIABLE
CMPLX     000000  COMPLEX*8 PROCEDURE
X         000024  REAL*4    VARIABLE
I         000030  INTEGER*2 VARIABLE
COS       000000  REAL*4    PROCEDURE
SIN       000000  REAL*4    PROCEDURE
```

```
      C --- SUBROUTINE TO GENERATE LINE SHAPE
      C --- LTYP DETERMINES SHAPE:
      C     0 --> READ SHAPE ARRAY (UNIT 2)
      C     1 --> GENERATE GAUSSIAN SHAPE
      C     2 --> GENERATE LORENTZIAN SHAPE
      C --- STORE SHAPE IN X-ARRAY
      C --- STORE # POINTS IN NX
      C --- WIDTH IS FWHM OF DESIRED LINE FOR LTYP = 1,2
      C --- FMT IS RUN TIME FORMAT
0001        SUBROUTINE GENER(LTYP,X,NX,FWHM,IX)
0002        DIMENSION X(200),FMT(20)
0003        IF(LTYP.EQ.1) GO TO 20
0005        IF(LTYP.EQ.2) GO TO 30
      C --- READ IN SHAPE
0007        READ(2,100) NX
0008  100   FORMAT(I5)
0009        READ(2,101) FMT
0010  101   FORMAT(20A4)
0011        READ(5,FMT) (X(I),I=1,NX)
0012        GO TO 40
      C --- GENERATE GAUSSIAN
0013  20    T=4.*ALOG(2.)/FWHM**2
0014        KF=SQRT(7./T)
0015        NX=KF*2+1
0016        KF=KF+1
0017        DO 21 I=1,NX
0018  21    X(I)=EXP(-(I-KF)**2*T)
0019        GO TO 40
      C --- GENERATE LORENTZIAN
0020  30    T=FWHM/2.
0021        KF=30.*T
0022        NX=KF*2+1
0023        KF=KF+1
0024        DO 31 I=1,NX
0025        D=I-KF
0026  31    X(I)=T/(D**2+T**2)
      C --- NORMALIZE AREA
      C --- FIND MAXIMUM POINT, SET IX: X(IX)=MAX(X)
0027  40    S=0.
0028        IX=1
0029        DO 41 I=1,NX
0030        IF(X(IX).LT.X(I)) IX=I
0032  41    S=S+X(I)
0033        DO 42 I=1,NX
0034  42    X(I)=X(I)/S
0035        RETURN
0036        END
```

RT-11 FORTRAN IV   STORAGE MAP

NAME	OFFSET	ATTRIBUTES		
X	000016	REAL*4	PARAMETER ARRAY (200)	
FMT	000026	REAL*4	ARRAY (20)	
LTYP	000014	INTEGER*2	PARAMETER VARIABLE	
NX	000020	INTEGER*2	PARAMETER VARIABLE	
FWHM	000022	REAL*4	PARAMETER VARIABLE	
IX	000024	INTEGER*2	PARAMETER VARIABLE	
I	000170	INTEGER*2	VARIABLE	
T	000172	REAL*4	VARIABLE	
ALOG	000000	REAL*4	PROCEDURE	
KF	000176	INTEGER*2	VARIABLE	
SQRT	000000	REAL*4	PROCEDURE	
EXP	000000	REAL*4	PROCEDURE	
D	000200	REAL*4	VARIABLE	
S	000204	REAL*4	VARIABLE	

RT-11 FORTRAN IV   V01-11   SOURCE LISTING   PAGE 001

```
        C --- GENERATE GAUSSIAN DISTRUBITION
0001            FUNCTION GAUS(IR1,IR2)
        C --- AVERAGE 15 SAMPLES OF EVEN DISTRIBUTION
        C --- MEAN = 1/2
        C --- VARIANCE = 1/12
0002            S=0.
0003            DO 1 I=1,15
0004    1       S=S+RAN(IR1,IR2)
0005            GAUS=S/15.
0006            RETURN
0007            END
```

RT-11 FORTRAN IV   STORAGE MAP

NAME	OFFSET	ATTRIBUTES	
GAUS	000020	REAL*4	VARIABLE
IR1	000014	INTEGER*2	PARAMETER VARIABLE
IR2	000016	INTEGER*2	PARAMETER VARIABLE
S	000024	REAL*4	VARIABLE
I	000030	INTEGER*2	VARIABLE
RAN	000000	REAL*4	PROCEDURE

```
        C --- SMOOTH.FOR
        C --- SECTINAL QUADRITIC LEAST-SQUARES FIT
        C --- FOR NS > 0, SMOOTH A INTO B
        C --- NS SHOULD BE < 25.
0001            SUBROUTINE SMOOTH(A,ND,NS,B)
0002            DIMENSION A(ND),B(ND),S(49)
0003            IF(NS.LE.0) RETURN
        C --- CALCULATE SMOOTHING PROFILE
0005            NSS=2*NS+1
0006            NS1=NS+1
0007            R=NSS
0008            A2=NS*NS1*R/6.
0009            AA=3.*NS*NS1-1.
0010            XNORM=R*AA-10.*A2
0011            DO 1 I=1,NS
0012            R=I
0013            P=(AA-5.*R*R)/XNORM
0014            S(NS1+I)=P
0015    1       S(NS1-I)=P
0016            S(NS1)=AA/XNORM
        C --- APPLY SMOOTHING
0017            CALL CONV(A,ND,S,NSS,NS1,B,1)
0018            RETURN
0019            END
```

RT-11 FORTRAN IV   STORAGE MAP

NAME	OFFSET	ATTRIBUTES	
A	000014	REAL*4	PARAMETER ARRAY (ND)
ND	000016	INTEGER*2	PARAMETER VARIABLE
B	000022	REAL*4	PARAMETER ARRAY (ND)
S	000024	REAL*4	ARRAY (49)
NS	000020	INTEGER*2	PARAMETER VARIABLE
NSS	000332	INTEGER*2	VARIABLE
NS1	000334	INTEGER*2	VARIABLE
R	000336	REAL*4	VARIABLE
A2	000342	REAL*4	VARIABLE
AA	000346	REAL*4	VARIABLE
XNORM	000352	REAL*4	VARIABLE
I	000356	INTEGER*2	VARIABLE
P	000360	REAL*4	VARIABLE
CONV	000000	REAL*4	PROCEDURE
*			

RT-11 FORTRAN IV   V01-11 SOURCE LISTING   PAGE 001

```
     C --- WEIGHT FUNCTION
0001         FUNCTION WT(KW,X)
     C --- KW:
     C     0 NO WEIGHT (WT=1)
     C     1 IMPOSE 0-->1 LIMIT WITH WT=X*(1-X)
     C     2 IMPOSE 0 LOWER LIMIT WITH WT=X,X<1, WT=1,X>1
0002         WT=1
0003         GO TO (1,2),KW
0004         RETURN
0005  1      WT=X*(1.-X)
0006         RETURN
0007  2      IF(X.GT.1.) RETURN
0009         WT=X
0010         RETURN
0011         END
```

RT-11 FORTRAN IV   STORAGE MAP

NAME   OFFSET ATTRIBUTES

NAME	OFFSET	ATTRIBUTES		
WT	000020	REAL*4	VARIABLE	
KW	000014	INTEGER*2	PARAMETER	VARIABLE
X	000016	REAL*4	PARAMETER	VARIABLE

USING THE TEST PROGRAM LONGO

The FORTRAN program LONGO, which is listed in Appendix 4, is used to deconvolve data files that are too large to be fit into our computer's limited memory. Our data files are written in two byte words (0 to 32767), 250 for each unformatted record. Two records are kept in memory at any one time, in two arrays for each of three files. The data file to be deconvolved is associated with unit 1 and is stored in IX1 and IX2, and the current deconvolution result and the result of the last iteration are associated with units 2 and 3 and are stored in IZ1, IZ2, IY1, and IY2. In each deconvolution iteration, when the end of the second array is reached, IZ1 (the current result) is written out, IX2 is copied into IX1, IY2 into IY1 and IZ2 to IZ1, and the next record is read into IX2 and IY2 from the data file and last iteration file, respectively. At the end of each iteration, the files are rewound and the roles of units 2 and 3 are reversed by changing the unit numbers associated with the data files.

The response function is read from unit 4 and is in the same format as the data files except that $a - 1$ marks the end of data. After the response function is read in, it is normalized to unit area. Unit 5 is the control or command file from which the control parameters are read. Unit 7 is the message file and is always the teletype. Unit 8 is associated with a log file where the deconvolution parameters are echoed and the root-mean-square error is written for each iteration.

The deconvolution parameters are LOOP, the number of iterations to perform with the given parameters, ALF, which is just α_{max}; KWT, the weight code; and KINT, the interpolation interval. ALF, KWT, and KINT are the same as ALF, KW, and KK of the program DECO. The format for the control parameters is (I5, F5.0, I5, I5). After LOOP iterations are performed, another set of control parameters are read until LOOP = 0 is detected, when the program exits.

The data files associated with units 1, 2, and 3 need to be explained further. The file associated with unit 3 is a temporary file and is deleted on program exit. Units 1 and 2 are permanent files, with unit 1 always associated with the data file to be deconvolved, which is not altered, and with unit 2 associated with either the current result or the result of the last iteration. On the first iteration, unit 2 is read in and deconvolved; the result is stored in unit 3, one record at a time. On the second iteration, the roles of the two files are reversed by exchanging the variable unit numbers in lines 92 through 94. After an even number of iterations, the result is stored back in the permanent file associated with unit 2. When the program exits, if an odd number of iterations have been performed, the temporary file is copied into the permanent file. This procedure should work for most versions of FORTRAN. If not, it is recommended that the temporary file be copied at the end of each iteration. The file associated with unit 2, the deconvolved result, can be initialized by copying the data file to be deconvolved. This also allows one to rerun the program and begin deconvolving with a previous deconvolution result if it is desired to perform further deconvolution iterations.

LONGO

RT-11 FORTRAN IV   V01-11 SOURCE LISTING

```
          C --- LONGO.FOR
          C --- LONG SCAN DECONVOLUTOR
          C --- UNIT 1 IS FILE TO BE DECONVOLUTED
          C --- UNITS 2 AND 3 ARE THE CURRENT AND LAST ITERATIONS
          C --- UNIT 2 IS AN EXISTING FILE AND WILL CONTAIN THE LAST
          C     ITERATION ON PROGRAM EXIT
          C --- UNIT 3 IS A TEMPORARY FILE
          C --- UNIT 4 CONTAINS THE RESPONSE FUNCTION
          C --- UNIT 5 IS THE COMMAND OR CONTROL FILE
          C --- UNIT 7 IS THE MESSAGE FILE (TT:)
          C
          C --- COMMAND FILE CONTAINS (I5,F5.0,I5,I5) :
          C           LOOP,ALF,KWT,KINT    WHERE
          C           LOOP IS THE # OF ITERATIONS TO PERFORM, 0-> EXIT
          C           ALF IS ALPHA, THE CORRECTION WEIGHT
          C           KWT IS THE WEIGHT CODE (SEE FUNCTION WT)
          C           KINT IS THE INTERPOLATION INTERVAL
          C
0001            DIMENSION IX1(250),IY1(250),IZ1(250),IX2(250),IY2(250),
          *        IZ2(250),IX(500),IY(500),IZ(500),F(250),G(251),FF(250)
0002            EQUIVALENCE (IX(1),IX1(1)),(IX(251),IX2(1))
0003            EQUIVALENCE (IY(1),IY1(1)),(IY(251),IY2(1))
0004            EQUIVALENCE (IZ(1),IZ1(1)),(IZ(251),IZ2(1))
0005            CALL ASSIGN(7,'TT:')
0006            WRITE(7,300)
0007            CALL ASSIGN(5,,-1,'RDO')
0008            WRITE(7,299)
0009            CALL ASSIGN(8,,-1,'NEW')
0010            WRITE(7,301)
0011            CALL ASSIGN(1,,-1,'RDO')
0012            WRITE(7,302)
0013            CALL ASSIGN(2,,-1,'OLD')
0014            WRITE(7,303)
0015            CALL ASSIGN(3,,-1,'TMP')
          C --- READ IN RESPONSE FUNCTION & NORMALIZE, -1 MARKS END
          C --- SET NF TO # OF POINTS
          C --- SET NF1 TO MAXIMUM POINT
0016            WRITE(7,304)
0017            CALL ASSIGN(4,,-1,'RDO')
0018            READ(4) IX1
0019            DO 10 I=1,250
0020            IF(IX1(I).LT.0) GO TO 12
0022      10    S=S+IX1(I)
0023      12    NF=I-1
0024            DO 11 I=1,NF
0025      11    F(I)=IX1(I)/S
0026            S=F(1)
0027            DO 13 I=1,NF
0028            IF(F(I).LE.S) GO TO 13
0030            NF1=I
0031            S=F(I)
0032      13    CONTINUE
          C --- SET UNIT POINTERS
0033            L1=2
0034            L2=3
0035      40    READ(5,500,END=77) LOOP,ALF,KWT,KINT
0036            IF(LOOP.EQ.0) GO TO 77
```

```
0038            WRITE(8,502) LOOP,ALF,KWT,KINT
0039            XINT=KINT
0040            NFF=NF/KINT
0041            KKK=KINT-1
0042            DO 30 L=1,LOOP
0043            READ(1) IX1
0044            READ(1) IX2
0045            READ(L1) IY1
0046            READ(L1) IY2
0047            R=0.
0048            DO 19 I=1,NF
0049     19     IZ(I)=0
         C --- LOAD G ARRAY WITH EVERY KINT-TH DATA POINT
         C --- AND FF ARRAY WITH EVERY KINT-TH POINT OF F, AND RENORMALIZE
0050            S=0.
0051            DO 20 I=1,NFF
0052            G(I)=IY(I*KINT)
0053            FF(I)=F(I*KINT)
0054     20     S=S+FF(I)
0055            DO 220 I=1,NFF
0056     220    FF(I)=FF(I)/S
0057            SD=0.
0058            K=NF1+(L-1)-(L-1)/KINT*KINT
0059            M=NFF*KINT
0060     22     M=M+KINT
0061     222    IF(M.GT.500) GO TO 50
0063            G(NFF+1)=IY(M)
0064            S=0.
0065            DO 21 I=1,NFF
0066            S=S+FF(I)*G(I)
0067     21     G(I)=G(I+1)
0068            D=IX(K)-S
0069            R=R+1.
0070            SD=SD+D*D
         C --- CALCULATE UMWEIGHTED CORRECTION
0071            D=D*WT(KWT,IY(K)/32767.)*ALF
0072            IF(KINT.EQ.1) GO TO 24
         C --- APPLY INTERPOLATED CORRECTION IF KINT > 1
0074            DO 23 I=1,KKK
0075            C=D*(KINT-I)/XINT
0076            IZ(K-I)=IY(K-I)+C
0077     23     IZ(K+I)=IY(K+I)+C
0078     24     IZ(K)=IY(K)+D
         C --- INCREMENT K
0079            K=K+KINT
0080            GO TO 22
         C --- RECIRCULATE BUFFERS
0081     50     WRITE(L2) IZ1
0082            DO 51 I=1,250
0083            IX1(I)=IX2(I)
0084            IY1(I)=IY2(I)
0085     51     IZ1(I)=IZ2(I)
0086            READ(1,END=86) IX2
0087            READ(L1) IY2
0088            M=M-250
0089            K=K-250
0090            GO TO 222
```

```
0091   88      WRITE(L2) IZ2
0092           LL=L2
0093           L2=L1
       C --- RECIRCULATE UNIT POINTERS AND RESET FILES
0094           L1=LL
0095           REWIND 1
0096           REWIND 2
0097           REWIND 3
0098           SD=SQRT(SD/R)
0099   30      WRITE(8,501) L,SD
0100           GO TO 40
0101   77      IF(L2.EQ.3) GO TO 78
0102   79      READ(3,END=78) IZ1
0103           WRITE(2) IZ1
0104           GO TO 79
0106   78      CALL EXIT
0107   299     FORMAT('$LOG FILE ')
0108   300     FORMAT('$COMMAND FILE ')
0109   301     FORMAT('$DATA FILE ')
0110   302     FORMAT('$DECONVOLUTED FILE ')
0111   303     FORMAT('$SCRATCH FILE ')
0112   304     FORMAT('$RESPONSE FUNCTION FILE ')
0113   500     FORMAT(I5,F5.0,I5,I5)
0114   501     FORMAT(' ITER = ',I3,'  RMSE = ',F8.0)
0115   502     FORMAT(' LOOP = ',I3,' ALF = ',F5.2,' KWT = ',I2,
       KINT = ',I2)
0116           END
```

```
RT-11 FORTRAN IV          STORAGE MAP

NAME     OFFSET   ATTRIBUTES

IX1      000006   INTEGER*2 ARRAY (250)
IY1      001756   INTEGER*2 ARRAY (250)
IZ1      003726   INTEGER*2 ARRAY (250)
IX2      000772   INTEGER*2 ARRAY (250)
IY2      002742   INTEGER*2 ARRAY (250)
IZ2      004712   INTEGER*2 ARRAY (250)
IX       000006   INTEGER*2 ARRAY (500)
IY       001756   INTEGER*2 ARRAY (500)
IZ       003726   INTEGER*2 ARRAY (500)
F        005676   REAL*4    ARRAY (250)
G        007646   REAL*4    ARRAY (251)
FF       011622   REAL*4    ARRAY (250)
ASSIGN   000000   REAL*4    PROCEDURE
I        014154   INTEGER*2 VARIABLE
S        014156   REAL*4    VARIABLE
NF       014162   INTEGER*2 VARIABLE
NF1      014164   INTEGER*2 VARIABLE
L1       014166   INTEGER*2 VARIABLE
L2       014170   INTEGER*2 VARIABLE
LOOP     014172   INTEGER*2 VARIABLE
ALF      014174   REAL*4    VARIABLE
KWT      014200   INTEGER*2 VARIABLE
KINT     014202   INTEGER*2 VARIABLE
XINT     014204   REAL*4    VARIABLE
NFF      014210   INTEGER*2 VARIABLE
KKK      014212   INTEGER*2 VARIABLE
L        014214   INTEGER*2 VARIABLE
R        014216   REAL*4    VARIABLE
SD       014222   REAL*4    VARIABLE
K        014226   INTEGER*2 VARIABLE
M        014230   INTEGER*2 VARIABLE
D        014232   REAL*4    VARIABLE
WT       000000   REAL*4    PROCEDURE
C        014236   REAL*4    VARIABLE
LL       014242   INTEGER*2 VARIABLE
SQRT     000000   REAL*4    PROCEDURE
EXIT     000000   REAL*4    PROCEDURE
```

```
0001          FUNCTION WT(KWT,X)
0002          WT=0.
0003          GO TO (1,2,3),KWT
0004          RETURN
0005    1     WT=1
0006          RETURN
0007    2     WT=X*(1.-X)
0008          RETURN
0009    3     WT=X
0010          IF(WT.GT..5) WT=.5
0012          RETURN
0013          END
```

RT-11 FORTRAN IV   STORAGE MAP

NAME	OFFSET	ATTRIBUTES	
WT	000020	REAL*4	VARIABLE
KWT	000014	INTEGER*2	PARAMETER VARIABLE
X	000016	REAL*4	PARAMETER VARIABLE

REFERENCES

1. W. E. Blass, *Appl. Spectrosc. Rev.* **11**, 57 (1976).
2. W. E. Blass and A. H. Nielsen, in *Molecular Physics, Part A* (D. Williams, ed.), Methods of Experimental Physics, 2nd ed., Vol. 3, pp. 126–202. Academic Press, New York, 1974.
3. J. E. Stewart, *Infrared Spectroscopy*. Dekker, New York, 1970.
4. E. A. Robinson, *Statistical Communication and Detection*. Hafner, New York, 1967.
5. E. A. Robinson, *Multichannel Time Series Analysis with Digital Computer Programs*. Holden-Day, San Francisco, California, 1967.
6. G. K. Wertheim, *J. Electron Spectrosc.* **6**, 239 (1975).
7. R. N. Jones, R. Verkataraghavan, and J. W. Hopkins, *Spectrochim. Acta, Part A* **23**, 925, 941 (1967).
8. P. A. Jansson, R. H. Hunt, and E. K. Plyler, *J. Opt. Soc. Am.* **60**, 596 (1970).
9. P. A. Jansson, *J. Opt. Soc. Am.* **60**, 184 (1970).
10. P. D. Willson, Ph.D. Thesis, Michigan State Univ., East Lansing, 1973.
11. P. A. Jansson, R. H. Hunt, and E. K. Plyler, *J. Opt. Soc. Am.* **58**, 1665 (1968).
12. B. R. Frieden, in *Picture Processing and Digital Filtering* (T. S. Huang, ed.), p. 179. Springer-Verlag, Berlin and New York, 1975.
13. J. Pliva, A. S. Pine, and P. D. Willson, *Appl. Opt.* **19**, 1833 (1980).
14. W. E. Blass, *Appl. Spectrosc.* **30**, 3 (1976).
15. J. A. Cadzow, *Discrete Time Systems*. Prentice-Hall, Englewood Cliffs, New Jersey, 1973.
16. B. B. Lathi, *Random Signals and Communication Theory*. International Textbook Co., Scranton, Pennsylvania, 1968.
17. L. E. Franks, *Signal Theory*. Prentice-Hall, Englewood Cliffs, New Jersey, 1969.
18. P. H. Van Cittert, *Z. Phys.* **69**, 298 (1931).

19. G. K. Wertheim, D. N. E. Buchanan, N. V. Smith, and M. M. Traum, *Phys. Lett. A* **49**, 191 (1974).
20. S. Hüfner and G. K. Wertheim, *Phys. Rev. B* **11**, 678 (1975).
21. G. Halsey and W. E. Blass, *Appl. Opt.* **16**, 286 (1977).
22. P. D. Willson and T. H. Edwards, *Appl. Spectrosc. Rev.* **12**, 1 (1976).
23. D. E. Jennings and W. E. Blass, *J. Mol. Spectrosc.* **55**, 445 (1975).
24. P. A. Jansson, *J. Opt. Soc. Am.* **60**, 184 (1970).
25. T. Todd and T. K. McCubbin, Jr., *Appl. Spectrosc.* **31**, 326 (1977).
26. A. Savitsky and M. J. E. Golay, *Anal. Chem.* **36**, 1627 (1964).
27. D. E. Jennings, Ph.D. Thesis, Univ. of Tennessee, Knoxville, 1974.
28. R. L. Burden, J. D. Faires, and A. C. Reynolds, *Numerical Analysis*. Prindle, Weber & Schmidt, Boston, Massachusetts, 1978.
29. A. Atakan, W. E. Blass, and D. E. Jennings, *Appl. Spectrosc.* **34**, 369 (1980).

BIBLIOGRAPHY

P. Areas and L. Hochard-Demolliere, Etude des intensités et des largeurs des raies de vibration-rotation de la bande deuxième harmonique de l'anhydride carbonique à 6970 cm^{-1} par déconvolution numeric, *Can. J. Phys.* **46**, 1967 (1968).

I. Balslev, J. E. Larsen, and S. Larsen, Noise amplification and resolution improvement in deconvolution of experimental spectra, *Appl. Spectrosc.* **32**, 454 (1978).

Y. Biraud, A new approach for increasing the resolving power by data processing, *Astron. Astrophys.* **1**, 124 (1969).

B. Carli, A discussion on data reduction in spectrometric measurements, *Infrared Phys.* **12**, 251 (1972).

G. Cesini, G. Guittari, G. Lucarini, and C. Pasma, An iterative method for restoring noisy images, *Opt. Acta* **25**, 501 (1978).

P. de Santis and F. Gori, On an iterative method for super-resolution, *Opt. Acta* **22**, 691 (1975).

B. R. Frieden, Restoring with maximum likelihood and maximum entropy, *J. Opt. Soc. Am.* **62**, 511 (1972).

F. J. Grunthaner, Electronic structure, surface reactivity and site analysis of transition metal complexes and metalloproteins by x-ray photoelectron spectroscopy, Ph.D. Thesis, Calif. Inst. Technol., Pasadena, 1974.

J. Guy, A. Salés, B. Brami-Depaux, and F. Joly-Cabaret, Notes des membres et correspondants et notes présentées on transmises par jeurs soins, *C. R. Acad. Sci., Ser. B* **265**, 109 (1967).

G. Halsey and W. E. Blass, Deconvolution of infrared spectra in real time, *Appl. Opt.* **16**, 286 (1977).

J. L. Harris, Diffraction and resolving power, *J. Opt. Soc. Am.* **54**, 931 (1964).

S. Hüfner and G. K. Wertheim, Core-line assymetries in the x-ray-photoemission spectra of metals, *Phys. Rev.* **11**, 678 (1975).

P. A. Jansson, Method for determining the response function of a high resolution infrared spectrometer, *J. Opt. Soc. Am.* **60**, 184 (1970).

P. A. Jansson, R. H. Hunt, and E. K. Plyler, Response function for spectral resolution enhancement, *J. Opt. Soc. Am.* **58**, 1665 (1968).

P. A. Jansson, R. H. Hunt, and E. K. Plyler, Resolution enhancement of spectra, *J. Opt. Soc. Am.* **60**, 596 (1970).

C. B. Johnson, Point-spread functions, line-spread functions, and edge response functions associated with MTFs of the form $\exp[-(w/w_c)^n]$, *Appl. Opt.* **12**, 1031 (1973).

N. Lacombe and A. Levy, A parametric deconvolution method: Application to two bands of N_2O in the 1.9 μm region, *J. Mol. Spectrosc.* **71**, 175 (1978).

D. P. MacAdam, Digital image restoration by constrained deconvolution, *J. Opt. Soc. Am.* **60**, 1617 (1970).

H. Sakai and G. Vanasse, Direct determination of the transfer function of an infrared spectrometer, *J. Opt. Soc. Am.* **56**, 357 (1966).

B. D. Saksena, K. C. Agarwal, D. R. Pahwa, and M. M. Pradhan, Convolution and deconvolution of Lorentzian bands, *Spectrochim. Acta, Part A* **24**, 1981 (1968).

G. K. Wertheim, Deconvolution and smoothing: Applications in ESCA, *J. Electron Spectrosc. Relat. Phenom.* **6**, 239 (1975).

H. Stone, Mathematical resolution of overlapping spectral lines, *J. Opt. Soc. Am.* **52**, 998 (1962).

G. K. Wertheim, Novel smoothing algorithm, *Rev. Sci. Instrum.* **46**, 1414 (1975).

G. K. Wertheim, D. N. E. Buchanan, N. V. Smith, and M. M. Traum, High resolution valence band spectra of the noble metals, *Phys. Lett. A* **49**, 191 (1974).

INDEX